Materials Science of Concrete:

SPECIAL VOLUME:

Calcium Hydroxide in Concrete

Related titles published by The American Ceramic Society:

Boing-Boing the Bionic Cat and the Jewel Thief
By Larry L. Hench
©2001, ISBN 1-57498-129-3

The Magic of Ceramics
By David W. Richerson
©2000, ISBN 1-57498-050-5

Boing-Boing the Bionic Cat
By Larry L. Hench
©2000, ISBN 1-57498-109-9

Ceramic Innovations in the 20th Century
Edited by John B. Wachtman Jr.
©1999, ISBN 1-57498-093-9

Materials Science of Concrete: Special Volume: Sulfate Attack Mechanisms
Edited by Jacques Marchand and Jan Skalny
©1999, ISBN 1-57498-074-2

Materials Science of Concrete: Special Volume: The Sidney Diamond Symposium
Edited by Menashi Cohen, Sidney Mindess, and Jan Skalny
©1998, ISBN 1-57498-072-6

Materials Science of Concrete V
Edited by Jan Skalny and Sidney Mindess
©1998, ISBN 1-57498-027-0

Materials Science of Concrete IV
Edited by Jan Skalny and Sidney Mindess
©1995, ISBN 0-944904-75-0

Materials Science of Concrete III
Edited by Jan Skalny
©1992, ISBN 0-944904-55-6

Materials Science of Concrete II
Edited by Jan Skalny and Sidney Mindess
©1991, ISBN 0-944904-37-8

Materials Science of Concrete I
Edited by Jan Skalny
©1989, ISBN 0-944904-01-7

For information on ordering titles published by The American Ceramic Society, or to request a publications catalog, please contact our Customer Service Department at 614-794-5890 (phone), 614-794-5892 (fax),<customer-srvc@acers.org> (e-mail), or write to Customer Service Department, 735 Ceramic Place, Westerville, OH 43081, USA.

Visit our on-line book catalog at <www.ceramics.org>.

Materials Science of Concrete:

SPECIAL VOLUME:

Calcium Hydroxide in Concrete

edited by

Jan Skalny • Juraj Gebauer • Ivan Odler

Published by
The American Ceramic Society
735 Ceramic Place
Westerville, Ohio 43081
www.ceramics.org

Proceedings of the Workshop on the Role of Calcium Hydroxide in Concrete, November 1–3, 2000, at Holmes Beach, Anna Maria Island, Florida.

Cover photo: "Secondary electron images showing the hexagonal habit of calcium hydroxite, needle-like habit of ettringite, and the sheet-like habit of calcium-silicate-hydrate," is courtesy of Paul E. Stutzman, and apperas as figure 1 in his paper "Scanning Electron Microscopy in Concrete Petrography." which begins on page 59.

Library of Congress Cataloging-in-Publication Data
A CIP record for this book is available from the Library of Congress.

ISBN: 978-1-57498-128-5

For information on ordering titles published by The American Ceramic Society, or to request a publications catalog, please call 614-794-5890.

4 3 2 1–04 03 02 01

Contents

FOREWORD

This special volume of The American Ceramic Society's Materials Science of Concrete series represents a collection or presentations given at the Workshop on the Role of Calcium Hydroxide in Concrete, held November 1–3, 2000, at Holmes Beach, Anna Maria Island, Florida.

The purpose of the industry-academia get-together was to discuss the physico-chemical role calcium hydroxide (also referred to as portlandite) plays in hydration and deterioration of cementing properties, as well as the implications of the presence of calcium hydroxide on the future of Portland, blended and specialty cements, and ecology of cement production.

Until now, calcium hydroxide, a major component of Portland clinker-based hydrated cements, has received little attention, notwithstanding the fact that many engineering properties of concrete are closely associated with its presence in hardened concrete. In almost all forms of chemical degradation of concrete, calcium hydroxide plays an important role and its excessive presence or absence may change the path of these reactions. Well-known examples of deterioration mechanisms involving calcium hydroxide are carbonation, corrosion of reinforcement, and alkali silica reaction, but also in other mechanisms, such as internal and external sulfate attack, calcium hydroxide plays an important role.

The workshop organizers and the editors of this special volume would like to acknowledge their appreciation to all workshop participants for their active involvement in the deliberations, with special thanks to the invited speakers. The Workshop would not have been possible without the sponsorship by numerous governmental and industrial organizations, specifically by National Institute of Standards and Technology and National Research Council of Canada, as well as by California Portland Cement Company, Essroc (Italcementi Group), Grace Construction Products, Holderbank Management and Consulting, Lafarge Canada, Materials Service Life, Portland Cement Association, RJ Lee Group, S.E.M., and Southdown.

We would also like to acknowledge the generous support by the Hon. Carol Whitmore, Mayor of Holmes Beach, as well as her staff, for allowing us to use the City Hall conference room and creating a pleasant working atmosphere. Thanks are also due to the reviewers of the submitted manuscripts for their thorough and timely reviews. Finally, in the name of all authors and participants, we would like to compliment the professionalism of the staff of The American Ceramic Society; their close cooperation with the editors enabled timely and quality publications of the Special Volume.

Jan Skalny
Juraj Gebauer
Ivan Odler

Organizing Committee

Jacques Marchand, Jan Skalny, and Paul Stutzman

Sponsoring Organizations

California Portland Cement Company, Glendora, CA, USA
Essroc – Italcementi Group, Middlebranch, OH, USA
Grace Construction Products, Cambridge, MA, USA
Holderbank Management & Consulting, Holderbank, Switzerland
Lafarge Canada, Montreal, QC, Canada
Materials Service Life, Pittsburgh, PA, USA
National Institute of Standards and Technology, Rockville, MD, USA
National Research Council, Ottawa, ON, Canada
Portland Cement Association, Skokie, IL, USA
R J Lee Group, Monroeville, PA, USA
S.E.M., Quebec, Canada
Southdown, Houston, TX, USA

Workshop Participants

Barger, Greg	Ash Grove Cement, KS
Beaudoin, James	National Research Council–Canada
Boyd, Andrew	University of Florida, FL
Chen, Hung	Southdown, TX
Clark, Boyd	Pennsylvania State University, PA
Delagrave, Anik	Ciment Lafarge–Canada
Diamond, Sidney	Purdue University, IN
Farkas, Emery	Consultant, FL
Figg, John	John Figg & Associates–U.K.
Gebauer, Juraj	Holderbank–Switzerland
Glasser, Frederick	University of Aberdeen–U.K.
Hawkins, Peter	California Portland Cement, CA
Hearn, Nataliya	University of Windsor–Canada
Hidalgo, Ana	Instituto Eduardo Torroja, Madrid–Spain
Jachimowicz, Felek	Grace Construction Products, MA
Johansen, Vagn	Construction Technology Laboratory, IL
Kinney, Fred	Essroc Materials, PA
Kirkpatrick, Jim	University of Illinois, IL
Marchand, Jacques	University of Laval–Canada
Mindess, Sidney	University of BC–Canada
Myers, David	Grace Construction Products, MA
Odler, Ivan	Professor Emeritus, Clausthal U.–Germany
Roy, Della	Pennsylvania State University, PA
Scrivener, Karen	Ciment Lafarge–France
Skalny, Jan	Materials Service Life, FL
Stark, Jochen	Bauhaus-Universitat, Weimar–Germany
Stutzman, Paul	U.S. Department of Interior–NIST, DC
Taylor, Hal	Professor Emeritus, U. of Aberdeen–U.K.
Tennis, Paul	Portland Cement Association, IL
Thaulow, Niels	R J Lee Group, PA
Thomas, Michael	University of Toronto–Canada
Turk, Danica	Consultant, IL

CALCIUM HYDROXIDE ISSUE: AN INDUSTRIAL VIEW

Juraj Gebauer

Holderbank Management and Consulting, CH-5113 Holdrbank, Switzerland

ABSTRACT

The presence of large quantities of calcium hydroxide in hardened concrete is a well-known and accepted fact. Such amounts are considered to be of importance for performance, particularly for faster hardening and protection of steel reinforcement. For more than a century now, the construction industry uses portland cement – mineral additive systems, and all attempts to modify or replace these systems by novel binders failed, primarily due to poor economic or technical performance, or both. During the long history of the existing binder systems we have learned quite successfully how to use and optimize them, as well as how to solve the occasional problems. The industry does not seem to be interested in changing the present situation in spite of some obvious advantages that could be realized if lesser amount of calcium hydroxide could be generated during the process of hydration.

Recognizing the conservatism of the industry and the slow pace of technical advances and technology transfer in this field, the author is still of the opinion that novel developments are possible and should be seriously considered by both academic and industrial communities. Such developments could result in more economic and more durable products.

INTRODUCTION

It is my pleasure and a great honor to be the first speaker at this special and unusual Workshop on concrete science. To the best of my knowledge, there was never a conference devoted to this very fundamental issue. And the location is also unusual indeed. For me, this is the preferred place to have a brain-storming session on the role of calcium hydroxide in concrete: a place where Nature shows good examples of calcium hydroxide-free -- high performance composites at the beaches of Anna Maria island.

To my surprise, I could find in the literature only relatively very few publications focusing on calcium hydroxide in concrete. The classical cement science focuses always on the hydration of portland or blended cements, with thousands of papers dealing with the composition and microstructure of hydrated cement reaction phases, without much attention to calcium hydroxide *per se*. Calcium hydroxide is considered to be a necessary "by-product" of cement hydration by most.

The need for calcium hydroxide in concrete is considered by most experts to be unquestionable. However, most of the discussion in the literature focuses only on the *quantity* of calcium hydroxide needed for good durability of steel imbedded in reinforced concrete; the positive effect of calcium hydroxide on the protection of steel reinforced concrete (control of pH) is the main reason for this view. The other, rather speculative advantage of calcium hydroxide is its protection of the C-S-H phase. It is also believed that only lime-rich cements render sufficiently high early strength of concrete, the most demanded property of concrete today.

The negative effects of calcium hydroxide on concrete properties are also obvious and acknowledged. Calcium hydroxide in concrete has a negative effect on its chemical resistance, leaching behavior, efflorescence, deterioration of concrete due to alkali-aggregate reaction, and durability in general. Calcium hydroxide probably enhances delayed ettringite formation in concrete as well. Its contribution to the strength performance of concrete is controversial. It is believed that coarse crystalline calcium hydroxide is disadvantageous for the strength development.

Overall, it seems that calcium hydroxide in concrete is recognized as a *necessary evil* and we have learned during the long history of portland cement concrete use concrete to live with this evil quite well. We have found acceptable solutions for most of the durability problems related to calcium hydroxide.

So is calcium hydroxide just a benevolent evil or is it the *Achilles' heel* -- the most vulnerable spot of the well-established giant, the portland cement concrete? That's what this Workshop is all about, I believe. Maybe other, completely new solutions should be examined to eliminate the earlier mentioned disadvantages of calcium hydroxide on concrete properties.

AN ATTEMPT TO FORMULATE AN INDUSTRIAL VIEW

In this part of my presentation, I would like to talk about the ever lasting and even widening gap between science/academia and industry, using the example of the calcium hydroxide issue.

Science has developed reasonably good knowledge on most aspects of concrete technology, at least sufficient enough to handle most of the practical problems confronting the industry. This knowledge is related to cement and concrete production practices, preventive measures for concrete constructions, diagnosis and repair, as well as the development of new types of binders. The industry, however, is using this knowledge only to a limited extent. The know-how transfer is lacking in efficiency. I believe that not using the available knowledge contributes yearly to multi-billion dollar losses to the society worldwide.

Why is the industry not using the existing knowledge more efficiently? The academic community is often frustrated and claims that industry is ignorant or not interested in any innovation and changes. I would like to elaborate on this topic from the industrial point of view and try to explain the industry's behavior.

In contrast to the science, where the driving force is curiosity and the objective is to generate new *knowledge*, industry is interested first of all in a successful *business*. The driving force is success on the world market. This means that economic feasibility and market acceptance of any new industrial products are essential. The road from an idea, an invention, to commercialization

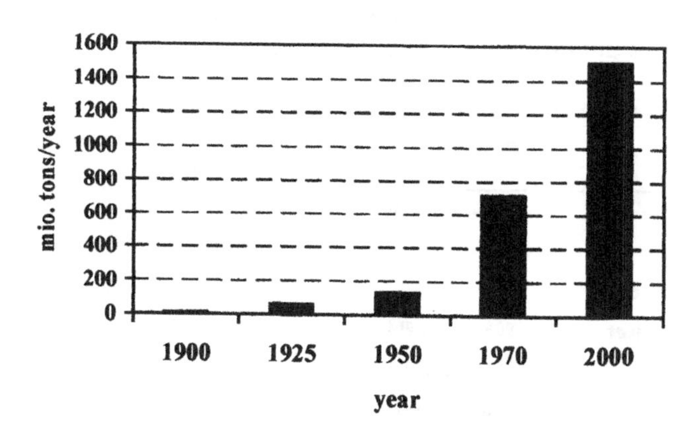

Fig. 1. Development of Cement production worldwide

is a very long, difficult, and expensive way, with many barriers and potholes. Most inventions fail due to these complex barriers. The many failures and their cost contribute to the diminished willingness of the cement and concrete industry to take risks.

During the last century, the cement industry has developed into a well organized and protectionist giant. The production rate rose from 10 million tons in 1900 to 1500 million tons in 2000 (Figure 1). This has significantly affected the business culture and makes any changes or revolutionary developments very difficult. The industry has evolved from small, often family owned companies demonstrating a certain pride with regards to their professional "guild", to huge multi-national companies where profit and market share decide the core business. At present, the industry is aiming to preserve the existing situation while cautiously exploring new opportunities. Substantial effort is given to optimization of portland and blended cement products by reducing production costs and improving or modifying the required properties. Having such a well-defined and established product, the industry, in essence, is a commodity business. Terms such as rationalization, automation, raw material resourcing, maintenance, logistics, economy of scale, shareholder value, and benchmarking are more important than the development of new products.

Numerous attempts to replace portland cement have failed thus far, mainly due to higher cost, but also due to poorer performance of the alternatives. Table I gives a selection of cement types that were invented, developed, and industrially produced since the introduction of the portland cement in 1824. It is interesting to note that with the exception of the dominating giant, the portland clinker based cements, all other cements show a lime content of less than 50% and, additionally, do not produce calcium hydroxide during cement hydration.

Table 1-a. Cement type - history

	Invention	Industrial production	Applications
Portland cement	1824 – J. Aspdin	1828 -	general - special
Blended cement	1876 – W. Michallis	1882 -	general - special
High alumina cement	1908 – J. Bied	1913 -	special
Supersulfated cement	1908 – H. Kühl	1932-1977	general – special
Alkali activated cement	1907 – H. Kühl	1965 -1990	general
Geopolymer cement	1976 – J. Davidovits	Not sign.	special
Sulfoaluminate cement	1932 – R. Pérré	1941-	Special (1 Mio.t/y)
Glass cement	1985 – J.F. Mc Dowel	Not sign.	special

Table 1-b. Cement type - chemistry

	CaO	Al_2O_3	SiO_2	$Ca(OH)_2$ in hydrated cement	Significant additional component
Portland cement	62-66	4-6	20-22	high	–
Blended cement	42-61	5-10	21-30	medium	–
High alumina cement	36-40	38-41	5-7	none	–
Supersulfated cement	43-45	11-12	26-28	none	sulfate
Alkali activated cement	40-42	11-12	30-32	none	alkalis
Geopolymer cement	11	18	60	none	alkalis
Sulfoaluminate cement	35-45	25-40	3-12	none	sulfate
Glass cement	42-48	28-35	16-22	none	–

Table 1-c. Cement type – cost and performance

	Production cost	Durability	Chemical resistance	CO2 emmision	Energy consumption
Portland cement	++	++	+	+++	+++(1450°C)
Blended cement	+?	+++	++	++	++
High alumina cement	+++	+?	+++	+	+++(1600°C)
Supersulfated cement	+	+++	+++	+	+
Alkali activated cement	+++	+++	+++	+	+
Geopolymer cement	++++	+++	+++	+	++
Sulfoaluminate cement	+++	++?	+++	+	++(1300°C)
Glass cement	++++	++?	+++	+	+++(1600°C)

+: low/no; ++++: very high; +?: questionable

Some of these cements were a commercial success, others are still struggling for market acceptance. A common feature of the non-portland cement is their good durability, and resistance against chemical attack, in particular sulfates. The protection of steel reinforcement against corrosion in some of the low lime cements seems to be secured. The examples given in the following show that presence of free calcium hydroxide is not required for the protection of steel in concrete.

The high alumina cement, developed and commercialized by Lafarge, and supersulfated cement, produced in Belgium, France, UK, and Germany from the 1930's to the 1970's, have a successful history. The production of the supersulfated cement was terminated in 1970's due to the deterioration of the available slag quality. The CaO/SiO_2 ratio of the slag decreased from 1.5 to 1.1 and it was impossible with the technology available at that time to produce good quality supersulfated cement.

The other cement types, such as alkali-activated slag, also known as soil cement, geopolymer cement and glass cement, are still in the development stage. The production of alkali-activated slag cement was initiated in Ukraine in 1965 and was terminated in 1990 after the societal changes in the former Soviet Union, primarily due to the lack of building construction activities. The expansive and fast setting sulfo-alumino cements have only a limited record of success in the field of sophisticated and special applications. The total amount of sulfo-alumino cement produced is very small compared to that of portland cement (~0.1%).

Despite of the well-known conservatism and skepticism of the cement industry, I personally believe that the industry will eventually respond and support the development of new cement types *providing that the economic feasibility and market acceptance are achieved.* This is something our academic friends need to remember. Collaborative efforts between the academic community and industry may help to change or modify this attitude of the industry -- from the conservative approach of preserving the existing cement systems to a progressive approach of supporting the development and commercialization of new cement types.

FIELD EXPERIENCE WITH "LOW LIME" CEMENTS

An example of good durability of supersulfated cement is that of the Beervlei Dam built in South Africa during 1954-1956 (Figure 2). After more than 40 years of service, the concrete has shown no signs of deterioration and no corrosion of the steel reinforcement has been observed. The measured concrete strength on the tested cores was about 124 MPa.

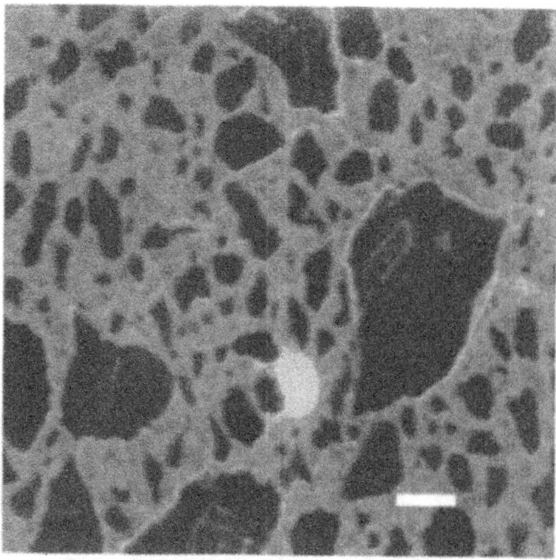

Figure 2a - Beervlei Dam in South Africa, built with supersulfated cement during 1954-1956. 2b. Microstructure of a concrete core taken from Beervlei dam in 2000.

The hydrated cement paste in the Beervlei concrete consists of dense, fine grained ettringite embedded in a "gel" of alumina and C-S-H. No calcium hydroxide was identified. Carbonation occurred at the surface (2-5 mm thick) of the concrete and showed a somewhat different phase composition, predominately calcite, gypsum, "gel" of alumina, and C-S-H. The microstructure was found to be sound and very dense also in this carbonated layer.

Similar good results were obtained on the 12-25 years old concretes produced from alkali activated slag in Ukraine (Figure 3). Alkali activated slag concrete was used in high rise buildings (Lipetsk) and in agricultural and industrial structures. The steel reinforcement of most concrete structures was not corroded and showed extremely dense amorphous microstructures of the hydrated cement with some microcracks. It is interesting to note that for both supersulfated and alkali activated cements no expansion with reactive aggregates is seen in the NBRI test (Figure 4). As these hydrated cements do not contain calcium hydroxide, it is assumed that for the expansion reaction the presence of calcium hydroxide is essential.

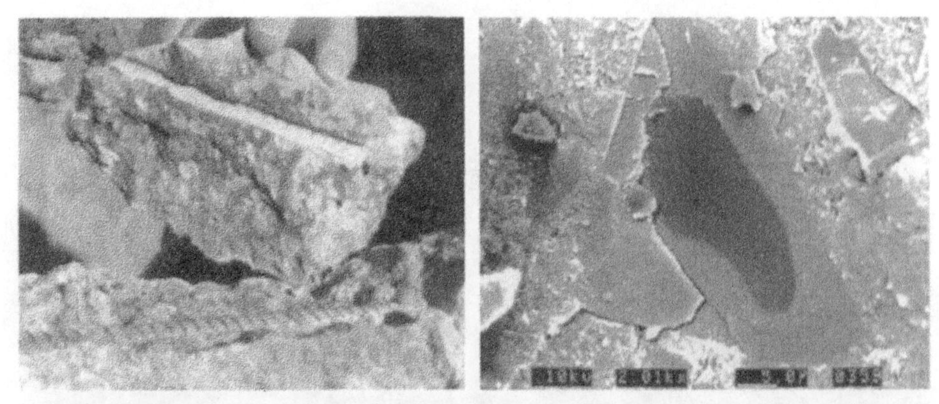

Fig. 3a. Concrete from a high rise building in Lipetsk, Ukraine;structure built with alkali activated slag cement during 1988. 3b. Photomicrograph shows the hydrated slag in the concrete core.

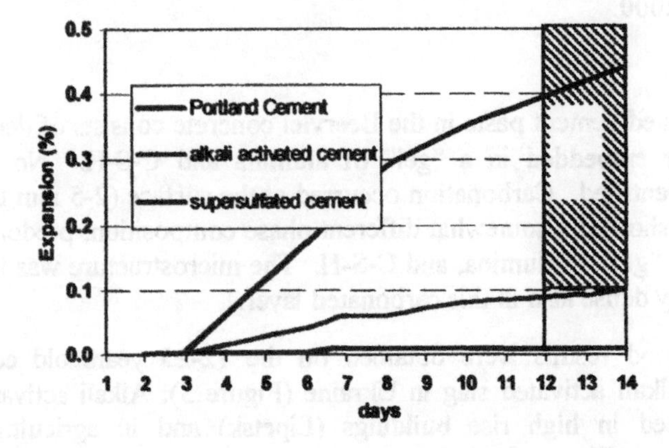

Fig. 4 - Comparison of portland, supersulfated and alkali activated cements with respect to their alkali-aggregate reactivity.

ENVIRONMENTAL ASPECTS

In the last part of my presentation, I would like to address the subject of the affect of environmental constraints on the future of our industry.

The increasing cost of the environmental protection – the cost for land filling or disposal of industrial and municipal by-products (wastes), cost of energy, possible CO_2 taxes, availability and cost of natural raw materials -- will force our industry to consider further minimizing the use of natural resources, decreasing the energy consumption, and disposal of wastes generated in the production processes. In particular, the industry will have to further minimize the use of limestone and decrease thermal energy consumption.

The overall goal is maximum recycling of materials within the production processes themselves, with minimum rejects. Environmental protection is actually a new business opportunity for the cement industry. In some regions this idea has received already reasonable attention and financial gain through the use of alternative raw materials and fuels for the cement production. Recycling of by-products from other industries, such metallurgical industry and thermal power plants, is becoming routine and will certainly increase in the future.

At the present there is practically no co-operation or joint venture between the major industrial branches such as cement, steel, and electricity producers. Such co-operation may happen in the future in order to maximize the synergies. The simultaneous production of cement and electricity or cement and steel sounds as Utopia today, but in the future it can become realistic. There are already available ideas and proposal in this direction. Furthermore, the recycling of old concrete to new products – including cement and aggregate -- could have in the future economic and ecological justification.

These futuristic thoughts are only very loosely related to our conference topic; nevertheless, I believe, for our general discussion this aspect of the calcium hydroxide issue may lead to some new insight.

CONCLUDING REMARKS

Finally, in conclusion I would like to return to what I have mentioned in the Introduction. Nature produces excellent calcium hydroxide-free construction material exemplified by seashells. During my previous stay at the Anna Maria island, I have collected seashells from the beach and analyzed them. In Figure 5, the microscopic features and XRD diagram of a seashell are shown. You may recognize the favorable layered structure of the pure aragonite-based composite material.

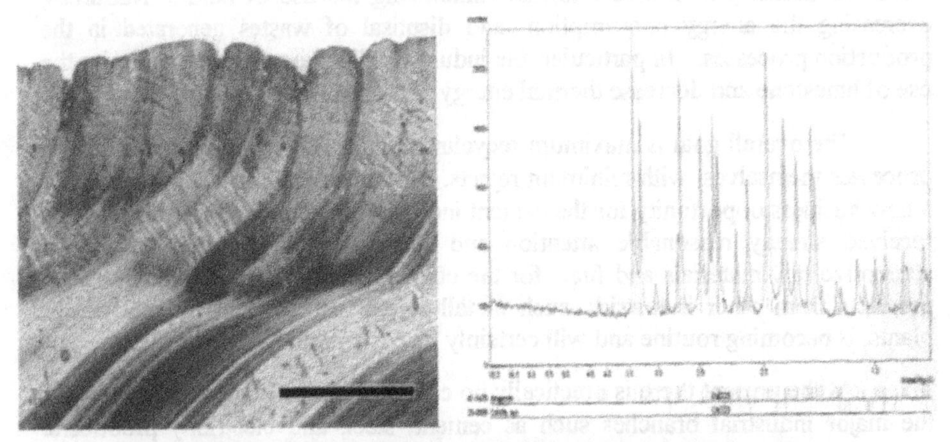

Fig. 5a. Photomicrograph and 5b. XRD pattern of a seashell.

This example may challenge us to copy the Nature better and to develop a calcium hydroxide-free cement-based solid the properties of which may be closer to the excellent properties of the superior models of the Nature, in this case the seashell.

One of the obvious challenges we face with respect to the development of new types of cements, including low-lime cements, is the availability of appropriate raw materials. Close cooperation between diverse industries for better utilization of industrial by-products is one of the many possible avenues to explore.

I'm confident that during the Workshop we shall have enough time to discuss the outlined views and ask new questions. My intention was to challenge the participants and to stimulate open discussions on these, for the cement industry essential, issues. Thank you for your attention.

THE ROLE OF $Ca(OH)_2$ IN PORTLAND CEMENT CONCRETES

F. P. Glasser
University of Aberdeen,
Chemistry Department,
Meston Building,
Meston Walk,
Aberdeen, AB24 3UE

INTRODUCTION

For millennia, cements based on $Ca(OH)_2$ were used as joining materials for brick and stone. $Ca(OH)_2$ itself is rather water soluble, *ca* 1.1 g/*l* at 20°C, so it is fortunate that the rather permeable sand-lime mortars used in historic structures carbonate readily; the lower solubility of $CaCO_3$, relative to $Ca(OH)_2$, helps improve resistance of the cement to dissolution. However even ancient civilisations were aware that the strength and durability of $Ca(OH)_2$ mortars could be further improved. Broadly, two methods were used; first, the use of impure limestone meals, which developed a pozzolan in the course of heating, and second, blending $Ca(OH)_2$ with pozzolanic materials, e.g. siliceous volcanic ash, prior to use.

The development of modern Portland cements represented a major materials advance over $Ca(OH)_2$ cements. Portland cement can, like $Ca(OH)_2$, be used in mortars and concretes but the resulting products have higher strengths, are less permeable and more resistant to environmental degradation. Furthermore, if necessary, they will set and harden under water, hence the term "hydraulic cement" to distinguish Portland (and other) cement types from $Ca(OH)_2$ - based materials.

The hydration of Portland cement pastes also yields $Ca(OH)_2$, amongst other solids, thereby providing continuity with past practise. Moreover, mixed $Ca(OH)_2$ - Portland cement formulations were (and still are) widely used in formulating mortars. Such formulations contain two sources of free $Ca(OH_2)$: from Portland cement (PC), as well as from the content of admixed hydrated lime, i.e., $Ca(OH)_2$. However hydrated lime- PC mixes will be excluded from this presentation.

The composition and mineralogy of Portland cement have not remained constant but have evolved with time. One evolutionary development has been an increase in lime saturation factor, LSF. Partly because the LSF concept was introduced in the 20th century, it is difficult to follow its numerical evolution with time. It is probable that meso-Portland cements of the mid 19th century had relatively low LSF's, perhaps 2.2 to 2.5, whereas in modern production LSF's are above 2.5 and are often close to 3.0. Back-of-envelope calculations discloses that this increase, although not large numerically, tends substantially to increase the amount of free $Ca(OH)_2$ which will form in a fully hydrated paste: I return to this subsequently.

Portland cement is of course mainly used to make concrete, a composite consisting of cement, sand, coarse aggregate and water. There is a general but ill-defined feeling that modern concrete is less durable than concrete made, say, 100 years ago. Certainly, modern concrete can be much stronger than concrete made 100 years ago: the improvement in early and late strengths is due to use of lower water : cement ratios, adding plasticisers if necessary to control fluidity, and finer grinding of cement. Durability is a more subjective term and, moreover, is difficult to quantify from present-day properties. But if we were to operate on the premise that the durability of modern Portland cement has been compromised and that its $Ca(OH)_2$ content is partly responsible, it ought not to be difficult to adduce evidence to support the assertion.

Much of the poor durability of Portland cement matrices, e.g. concrete, is in fact due to a combination of factors, all well-known and not associated with $Ca(OH)_2$ content. These include poor architectural design as well as a host of factors associated with the concrete itself: poor or inappropriate concrete mix specification, inadequate cure and compaction and poor workmanship and supervision. I must state at the outset that it is practicable to design and build durable structures using present-day Portland cements. At the same time, some service environments are relatively benign while others provide a more severe test of durability. Mixed modes of attack, in which concrete is exposed to several mechanisms of deterioration are an especially severe test of concrete quality. For example, a concrete pier in sea water is subject to attack by complex Na-Ca-Mg-Cl-SO_4 salt solutions, as well as by abrasion and, in the splash zone, to repeated wet-dry cycles, thermal fluctuations and possible freeze-thaw.

Nevertheless, $Ca(OH)_2$ is a major constituent of hydrated Portland cement; also many of the formulations specified for more durable concretes use blending agents - fly ash, slag, silica fume, metakaolin, etc. These have many impacts on the quality of the paste, one of which is to lower the free $Ca(OH)_2$ content.

My remit is to focus on $Ca(OH)_2$. This is appropriate because (i) in recent decades it has received comparatively little attention relative to other, less abundant constituent phases such as ettringite and (ii) because the life expectation of concrete, particularly as used in major infrastructural applications, e.g. roads, bridges, dams, is increasing. So it is appropriate to reassess the role of $Ca(OH)_2$. In order to set the stage for other presentations, I start at the beginning.

1. THE AMOUNT OF $Ca(OH)_2$ IN HYDRATED PORTLAND CEMENT

1.1 Methods

The amount of portlandite, $Ca(OH)_2$, in a paste can be determined by a number of techniques, some direct, others indirect. Table 1 lists the three most commonly-used methods together with comments on probable accuracy and limitations. Methods such as petrographic point counting and infrared have also been used. Point counting only detects optically-significant $Ca(OH)_2$ grains, > 1 to 2 microns: infrared appears to be semi-quantitative.

The writer is unaware of critical comparisons having been made between these methods using the same paste, but has experience of applying chemical extraction and thermal methods to the same pastes and finds that they generally agree to within ca 2% of the amount of $Ca(OH)_2$ present.

Calculations can also be used to determine the paste mineralogy, including portlandite content. Powers [1] presented models of cement hydration showing the content of pores, capillary water, gel water, gel solid and of residual cement. The "gel" was not subdivided into constituent solids but we can progress further as a result of modern investigations. I assume complete hydration, i.e., the alpha factor, as defined by Powers, is unity. The paste contains six chemical components: CaO, Al_2O_3, Fe_2O_3, SiO_2, SO_3 and H_2O. For purposes of calculation the minor components, e.g. MnO, P_2O_5, TiO_2, can be ignored and the six others recalculated to 100% as shown in Figure 1, step 1, Water in excess of that required to satisfy the hydration demands of the solids is assumed to be present. The remaining components are then divided amongst the phases. A number of arbitrary assumptions have to be made: for example, iron is arbitrarily excluded from the AFm and AFt structures but this has only minor impact on the distribution of silica and on calculated amounts of $Ca(OH)_2$ and C-S-H. If more refined calculations are desired, the scheme can be appropriately modified.

1.2 Comparison of methods and experiment

Figure 2 shows calculations for a simplified cement which are compared with experiment. Three scales are shown: two horizontal and one vertical. The horizontal scales show mole % $Ca(OH)_2$ in the bulk composition and C/S molar ratio, while the vertical scale shows mol % of portlandite, $Ca(OH)_2$. The three sloping lines are constructed according to differing assumptions: curve (a) assumes that the C-S-H has a Ca/Si molar ratio 2.0 while (b) and (c) assume ratios 1.8 and 1.5, respectively. Actual experimental measurements of the amount of $Ca(OH)_2$ were made by Hong [2] using the Franke extraction and TG methods, which gave consistent results and were in good agreement with calculations. The experiments used laboratory C-S-H made by reacting silica gel with $Ca(OH)_2$: the preparations were homogenised at 25° - 40°C for between 12 and 24 months. The actual experimental values lie on curve b until the Ca/Si ratio decreases below ~2, and thereafter follow curve d. Indeed, our reading of the literature is that most investigators who have prepared apparently homogeneous C-S-H find free portlandite in preparations made at C/S ratios > 1.5, although not in sufficient amounts to dispute the contention that C-S-H with a limiting C/S ratio close to 1.8 can be prepared, albeit with slight portlandite contamination. It does appear, however, that C-S-H with C/S > 1.5, approximately, is labile with respect to discharging $Ca(OH)_2$. To distinguish between the two types of calcium, free $Ca(OH)_2$ will be referred to as portlandite, as distinct from labile $Ca(OH)_2$ in C-S-H solid solution, whose actual form may not be known.

1.3 Labile $Ca(OH)_2$ in cement paste

Cement pastes differ from laboratory simulates: Table 2 highlights especially those relevant to the coexistence of portlandite and C-S-H. Lime-rich C-S-H is capable of contributing additional $Ca(OH)_2$ in appropriate circumstances.

This capability to donate calcium is well-known and is especially apparent in blended cement systems e.g. fly-ash-cement blends; with sufficient fly ash content, not only is free portlandite is consumed but the Ca/Si ratio of the C-S-H also decreases to within the range 1.2 to 1.5, thus contributing additional $Ca(OH)_2$. These pastes remain somewhat inhomogeneous for decades so the C-S-H composition may not be uniform: hence the large numerical range of C/S ratios. But the literature leaves no doubt that fly ash in sufficient quantity not only reacts with free portlandite but also removes "$Ca(OH)_2$" from the C-S-H phase.

There is other evidence that $Ca(OH)_2$ can be liberated from high-ratio C-S-H. Figure 3 summarises some of the observations: reactions with fly ash and silica fume have been noted: the C/S ratio can only be decreased by siliceous additives if mass balances are favourable: for fly ash, this implies high levels of replacement, perhaps 30 - 50% or more.

The isothermal annealing of commercial cement has been shown to lead to a different partition of calcium and silicon than is observed at 25°C. Paul and Glasser [3] compared the same OPC cured at 25°C and 85°C. After 8.4 years at 85°C they found that 16 - 25% siliceous hydrogarnet crystallised. Figure 4 shows the balances achieved amongst the hydrated solids. Much of the calcium necessary to form hydrogarnet came from C-S-H: while the C-S-H remained essentially non-crystalline, its C/S ratio decreased to ~1.5 even in the continuing presence of abundant free portlandite. Hong has shown that if laboratory preparations of C-S-H, initially with C/S close to 1.8, are annealed at temperatures in the range 40°C - 85°C, nanoscale segregation of $Ca(OH)_2$ occurs within weeks (85°C) or months (40°C). The crystallites of $Ca(OH)_2$ are too small to give an X-ray diffraction but are apparent by dark field electron microscopy and are confirmed as $Ca(OH)_2$ by their electron diffractions: a full description of this process is in the course of preparation.

It is thus apparent from a range of evidence that C-S-H in the composition range Ca/Si 1.5 to 2.0 is labile and can liberate $Ca(OH)_2$ in a range of situations. Thus C-S-H gels in the high-lime range become increasingly unstable and readily contribute $Ca(OH)_2$ to other, more acidic reactants, or will spontaneously exsolve portlandite upon prolonged gentle (40°C or more) heating within a few months. The boundary between persistent C-S-H gels and labile gels can only be fixed by operational criteria, i.e. the Ca-rich limit of C-S-H gel depends on the particular experimental methods used to generate and characterise labile $Ca(OH)_2$. But the limit of persistence appears to converge on C/S ratios in the range 1.5 as low as ~1.2. Thus the potential $Ca(OH)_2$ of a real cement could be viewed as consisting of two sources; one is measured by the actual free portlandite content while the other source is $Ca(OH)_2$ furnished by labile, high-ratio C-S-H. Reference to Figure 2 shows that for a simplified cement with LSF = 3.0, complete hydration would yield ~24% portlandite assuming the C-S-H phase has Ca/Si = 2.0. But if this ratio were to decrease to 1.5, the anticipated content of $Ca(OH)_2$ would increase to ~37%. This increased total would of course include portlandite as well as labile $Ca(OH)_2$.

The intrinsic state of labile $Ca(OH)_2$ in high ratio C-S-H can only be guessed. Two possibilities have been suggested in the literature and in discussion. One is that additional Ca ions are stuffed into the disordered C-S-H structure, the other, that more continuous protolayers of $Ca(OH)_2$, presumably on a nanoscale, are interstratified with tobermorite-like or jennite-like structural units. In this latter view, the structure of the C-S-H framework could resemble a poorly-ordered version of a mineral whose structure is known (tobermorite or jennite, although the jennite structure is conjectural). In this later view, the intercalated $Ca(OH)_2$ domains escape detection by X-ray diffraction and perhaps even detection by high resolution electron microscopy, at least until slightly coarsened by gentle heating.

2. MICROSTRUCTURE

It follows from the foregoing that $Ca(OH)_2$ is present, or may be present, in a hardened cement paste on a range of length scales: large, blocky crystals resulting from primary hydration and, possibly, structurally occluded, nanoscale $Ca(OH)_2$. Figure 5 shows schematically this range.

A common impression of $Ca(OH)_2$ microstructure is obtained from scanning electron micrographs of fracture surfaces. These disclose relative coarse, 1 - 10μm, blocky crystals and crystal aggregates. The perfect basal cleavage of $Ca(OH)_2$ is emphasised by fracture, which preferentially reveals surfaces normal to 00·1 (treating the crystals as having hexagonal symmetry). However, the true size distribution and morphology are not well revealed by fractographs which, as a consequence of cleavage, tend to divert fractures to pass through portlandite and thereby overestimate the relative volume occupied by $Ca(OH)_2$ in the matrix.

Another association of portlandite with embedded materials occurs at aggregate-paste interfaces [4]. In mortars and concretes, particularly in those made to high water : solid ratios, a margin of low-density paste surrounds, or partially surrounds, aggregate particles. This so-called 'process zone' has been estimated to range in thickness between a few microns to as much as 50 microns. The process zone is partially filled with portlandite, with perhaps minor and variable contents of C-S-H and ettringite. Similarly, in pastes which are reinforced with glass fibre, $Ca(OH)_2$ grows preferentially against the fibre, frequently to the extent that growth of $Ca(OH)_2$ platelets erodes and notch the fibre [5]. This is of course, destructive to fibre tensile strengths.

These associations correctly depict the microstructural relations of $Ca(OH)_2$ But, as shown, cement materials may contain a large size range of $Ca(OH)_2$ crystallite sizes. Thus while micron-sized $Ca(OH)_2$ is relatively well characterised, nanoscale $Ca(OH)_2$ remains an important yet incompletely explored constituent. Perhaps further studies by scattering (e.g. X-rays, neutrons) will contribute useful information on its content and distribution.

3. MECHANICAL PROPERTIES

It is tempting to speculate that the comparatively weak van der Waals bonds holding together electrically neutral sheets of portlandite constitute an intrinsic source of weakness.

The influence of $Ca(OH)_2$ on strength has been investigated by numerous workers Marchese [6] studied pastes made by hydrating alite, a chemically impure form of Ca_3SiO_5. This simulated the alite present in Portland cement. A paste made to w/c ratio 0.5 was hydrated for three years. Fractographs were analised using complimentary images, i.e. by locating the complimentary feature on both halves of a fracture, to determine if cracks passed through or around portlandite. Fracture was found to be cleavage-controlled: it passed through portlandite preferentially to the portlandite-C-S-H interfaces, suggesting that the C-S-H to portlandite bond was relatively strong. On the other hand, portlandite growing in pores was not fractured, suggesting that the pore itself was the weak feature. Thus the presence of portlandite, despite its perfect cleavage, was not viewed as a source of weakness in normal hardened cement pastes.

Mortars and concretes are more complex than paste, for example, another set of interfaces occurs between paste and aggregate. Struble [7] suggested that portlandite grew epitaxially on mineral aggregates. This was confirmed [8] and the *c* axis of portlandite was found preferentially to be normal to the interface. The zone of orientation may extend to as much as 50μm into the paste [9, 10]. While this might enable the cleavage to act as a strength-reducing factor, $Ca(OH)_2$ does occur mixed with other phases, e.g., ettringite. Also, the relatively poor crystallite packing in this process zone will itself act as a strength reducing flaw. Scrivener and Pratt [11] examined the microstructure of mortars ranging in age between 7d and 85 years and noted the occasional presence of "exceptionally large" crystals of $Ca(OH)_2$ in some older specimens. They tentatively suggested that formation of the larger crystals was conditioned by local variations in w/c ratio, high ratios allowing large portlandite crystals to grow. Thus large portlandite crystals were also associated with regions of low paste density.

It appears impossible on the basis of presently available evidence to deconvolute the various strength-limiting factors; there is no significant body of evidence to suggest that the strength of mortars or concretes is limited by the normal content of portlandite in modern cements. In this context it is noteworthy that high strength concretes, usually made to low w/c ratios, contain normal amounts of portlandite but its mean grain size is greatly reduced. However the content of large pores is similarly reduced. The rough proportionality between larger pore size and portlandite crystallite size contributes to the difficulty of deconvoluting the two factors. In this context it is also noteworthy that blended cement matrices, in which the amount of portlandite may be reduced, are not notably stronger than plain pastes if comparison is restricted to matrices having similar distributions of coarser porosity; also that C-S-H itself undoubtedly contains a range of local densities, as occur in the 'inner' and 'outer' product, as well as a range of hydrogen bond strengths, all of which could be sources of mechanical weakness. Thus there is no convincing evidence that the portlandite content of PC diminishes compressive strengths. Tensile strengths, which might be more sensitive indicators, appear not to have been investigated in this context.

$Ca(OH)_2$ AND THE DURABILITY OF CEMENT AND CONCRETE

Cement matrices experience a variety of degradation processes in a range of service environments. Space precludes a general development of degradation except insofar as the portlandite content is concerned and its possible special role. However, "degradation" needs qualification: most concretes contain embedded steel, and any processes which affect the permeation properties of cement paste also potentially affect the passivation of steel: Table 3 lists some of the processes which are relevant to degradation of cement solids.

Solubility data for portlandite in water are well known; at 18°C, its solubility is ~1.1 g/*l*. It is easy to generate supersaturated solutions, for example, by agitating an excess of $Ca(OH)_2$ in water, but solubilities quickly and spontaneously (*ca* 1 day) return to saturation. The dissolution process is of course congruent in the sense that each Ca^{2+} in solution is charge-balanced by $2OH^-$ ions. The pair $CaOH^+ \cdots OH^-$ may also be present but does not affect the congruent solubility. Portlandite is unusual inasmuch as its solubility *decreases* with increasing temperatures, at least up to 160° - 180°C. Thus the 100°C solubility is about 0.6g/*l*, approximately half that at 18°C.

The rather high solubility of portlandite in the ambient temperature range of engineered structures, say up to 40°C, suggests that calcium would be readily leached. There are, however, a range of circumstances which reduce the rate of calcium depletion relative to that which might be deduced from the solubility of portlandite in water: Table 4 summarises some important relevant factors.

Physical aspects of $Ca(OH)_2$ leaching are well described in the literature. For example, Carde, *et al*, [12] determined leach rates for $Ca(OH)_2$ from cement paste. They accelerated reaction using NH_4NO_3 solutions, but the results are said to simulate the leached zone developed in contact with distilled water. Leaching results in a decalcified zone marked by depletion of portlandite as well as decalcification of C-S-H, i.e., its C/S ratio decreases. The result of these changes was to increase both porosity and permeability of the leached layer and decrease its strength. In a companion paper, Delagave *et al* [13] showed that physical changes accompanying $Ca(OH)_2$ leaching could be modelled.

These studies appear to support the contention that portlandite contributes to enhanced degradation of concrete undergoing leaching. However, the real-life scenario is more often complex because other reactions occur simultaneously which affect calcium solubility and availability. Two are worthy of special mention: the presence of soluble alkali in cement or ground water, or both, and of CO_2.

Soluble alkali, effectively present in cement paste as NaOH and KOH, reduces the solubility of $Ca(OH)_2$ by the common ion effect. If the solubility of $Ca(OH)_2$ is written as a solubility product:

$$K_{SP} = [Ca]\,[OH]^2$$

Where brackets represent concentrations*: the hydroxide ions furnished by NaOH and KOH greatly reduce Ca solubility in order to maintain K_{SP}. From the shape of the solubility product expression, small amounts of NaOH and KOH have the greatest impact on reducing $Ca(OH)_2$ solubility. For this reason, a slow outward flux of alkali ions from cement to leachant may be effective in reducing calcium solubility for long periods of time. The depression of soluble calcium affects portlandite solubility as well as Ca leaching from high-ratio C-S-H.

A second important factor is that concrete is often leached by contact with rain water or fresh water in streams, lakes, rivers, etc., and such waters are normally saturated with respect to CO_2. Portlandite reacts strongly with dissolved aqueous CO_2 forming $CaCO_3$, the solubility of which is several orders of magnitude less than that of $Ca(OH)_2$; the relevant solubility products are 0.99×10^{-8} ($CaCO_3$, 18°C) and 9.95×10^{-4} ($Ca(OH)_2$, 20°C). Of course solubility products do not tell the whole story, but the greatly reduced solubility of $CaCO_3$ relative to $Ca(OH)_2$ tends to establish a semi-protective skin of calcium carbonate on cement paste. Silica gel, derived from C-S-H, contributes to this skin. The carbonation mechanism of $Ca(OH)_2$ has been

* The treatment is simplistic: species activities need to be taken into account. However the simple equation, using molar concentrations, is nevertheless a good approximation for leaching into initially pure water and illustrates the principle involved.

shown to involve expitaxial formation of a thickening layer of $CaCO_3$ on $Ca(OH)_2$. The $CaCO_3$ product is polycrystalline even when formed on a substrate of single-crystal portlandite [14]. This product layer is presumably microporous, although reaction slows with time, indicating that the product layer acts as a significant barrier to further reaction. C-S-H in cement seems to participate at approximately the same rate as portlandite. This may be a consequence of the paste microstructure and its significance to the overall kinetics is uncertain. Figure 6 is constructed to show the progress of carbonation in moist air. The cartoon shows how a diffuse carbonation front progresses, leaving behind islands of uncarbonated cement hydration product. The apparently sharp front revealed by the phenolpthalein test is misleading: the actual front is normally diffuse, not sharp. Of course, the carbonation reaction of C-S-H proceeds by a different mechanism than of portlandite: free silica, as amorphous hydrated silica, develops as a secondary product of carbonation together with $CaCO_3$, so the reacted product layer is inhomogeneous. The extent of hydration, and calcium content of the amorphous silica, are not well established; it is therefore difficult to calculate changes in specific volume attending carbonation of C-S-H. But for portlandite, the molar volume, $33cm^3$ per formula unit, increases to 36.9 cm^3upon conversion to calcite. Thus, despite the greater density of calcite relative to portlandite - 2.74 and $2.24g/cm^3$ respectively - carbonation is apparently slightly expansive. Considerable disagreement occurs in the literature about the volume changes and implications for porosity attending carbonation. In their review of the literature and from new measurements, Honda and Shizawa [15] report a decrease in porosity occurs during normal carbonation. However, while this may be true for an ideal situation, both dissolution and calcium migration from greater depths may accompany carbonation in saturated environments and the quantitative balance between factors is generally not known. It is therefore prudent to conclude that carbonation need not enhance porosity, but not to venture further without quantification of the mass and volume balances between rates of bulk migration, dissolution and carbonation.

In waters having a complex chemistry, e.g, sea water, reactions become more complex. Firstly, magnesium replaces Ca in both C-S-H and $Ca(OH)_2$: with $Ca(OH)_2$ the initial product of reaction is brucite, $Mg(OH)_2$. Sodium chloride solutions do not form a solid reaction product with portlandite or C-S-H at $<5M$, approximately, but it is often assumed that the presence of chloride will enhance the solubility of portlandite. Experiment does not provide evidence for significant solubility increases. Figure 7, taken from [16], shows the solubility of $Ca(OH)_2$ as a function of NaC*l* content and temperature. On any selected isotherm, 25° - 85°C, NaC*l* concentrations in the range to 1.5M increase solubility although the increase is numerically very small. Furthermore, the impact of increased temperature is to reduce solubilities at constant salinity.

Given that other components of sea water also react with $Ca(OH)_2$ and C-S-H, e.g. sulfate and dissolved CO_2, it is difficult to isolate specific chemistries which suggest that portlandite contributes disproportionately to the degradation of cement paste in sea water. Of course the balances between dissolution and precipitation are also pressure dependent: not only do sea water temperatures vary, but pressures vary with depth. Thus near-surface sea water is saturated with respect to $CaCO_3$ but becomes undersaturated at comparatively shallow depths. As a consequence, the solubility of $CaCO_3$ increases and its value as a protective layer against dissolution tends to decrease with depth; an analysis appropriate to shallow waters - which encompasses most structures - may not be appropriate for concrete in deep marine environments.

ADDITIONAL DISCUSSION

The presentation concentrates on reviewing only selected aspects of the role of $Ca(OH)_2$ in Portland cement. It is shown that Portland cement contains two sources of $Ca(OH)_2$; portlandite, occurring as physically discrete crystallites, and labile $Ca(OH)_2$. The latter is initially present as high Ca:Si ratio C-S-H gel but is (i) readily exsolved by gentle moist heating, (ii) available as an activator for pozzolanic reactions at all temperatures of interest and (iii) can become available for release during dissolution or carbonation reactions.

It is suggested that the metastability of C-S-H increases in the range Ca:Si ~1.5 to 2.0. This raises anew the possibility that the nanoscale structure of C-S-H consists of a basic unit, having Ca/Si ~1.5, containing intercalated $Ca(OH)_2$ in an X-ray amorphous form.

Evidence is also presented on the role of $Ca(OH)_2$ in controlling the strength of the paste. In theory, the easy cleavage of $Ca(OH)_2$ is mechanically similar to a flaw, in the Griffith's sense, of length equivalent to the size of the crystals. In practice, however, concrete contains so many other flaws of approximately the same length scale, or as in the case of paste-aggregate process zones, on a much larger scale, that weakening effects due to the presence of portlandite have not been convincingly demonstrated. High strength concretes made with plain Portland pastes are characterised by a reduction in mean portlandite crystallite size, so a decrease in cleavage-induced flaw size accompanies reduction is other flaws. Hence the gain in strength is achieved by changing several strength-limiting parameters simultaneously. In high-strength concretes, the two strength-controlling factors could be deconvoluted in principle by adding previously-formed coarsely crystalline $Ca(OH)_2$ to the formulation and comparing its mechanical properties to the equivalent formulation made without the additive.

The role of $Ca(OH)_2$ in deterioration processes is difficult to determine because several processes tend to occur simultaneously and no satisfactory method of deconvoluting the impact of these processes has yet been devised. The balance between chemical dissolution and species migration leads to formation of zoned structures. These are well characterised empirically: each zone represents a local equilibrium between cement paste and pore water, as modified by exchange with the local service environment. The thickness of zones varies with time, paste quality, intensity of attack and the extent to which dissolution weakens the matrix and enhances porosity and permeability of the affected regions. So many factors affect these processes that it is not possible to isolate the role of portlandite. However, just as equilibria in cement-groundwater interactions has been successfully modelled, we can anticipate that the increase in computing power, now available at comparatively low cost, will lead to development of combined kinetic-equilibria models, giving an expectation that the role of portlandite can be deconvoluted as a result of focussed efforts.

Other presentations in this Symposium will describe different aspects of the behaviour of $Ca(OH)_2$ in cement. It is apparent that much has been learnt but that much remains to be done in order to fully assess the role of $Ca(OH)_2$ in cement and concrete.

REFERENCES

[1] T. C. Powers. "The Physical Structure of Portland Cement Paste", Chapter 10 in "The Chemistry of Cements", (Ed. H. F. W. Taylor). (1964) Academic Press, London and New York.

[2] S.-Y. Hong. "Calcium Silicate Hydrate : Crystallisation and Alkali Sorption". PhD Thesis, University of Aberdeen (2000).

[3] M. Paul and F. P. Glasser. "Impact of Prolonged Warm (85°C) Moist Cure on Portland Cement Paste". Cement and Concr Res. (In press).

[4] "Interfaces in Cementitious Composites", (Ed., J. C. Maso). E. & F. Spon, London (1992).

[5] V. T. Yilmaz, E. E. Lachowski and F. P. Glasser. "Chemical and Microstructural Changes at Alkali-Resistant Glass Fibre - Cement Interfaces". Jour Amer. Ceram. Soc. **74** No. 12 3054 - 3060 (1991).

[6] B. Marchese. SEM Topography of Twin Fracture Surfaces of Alite Pastes 3 Years Old. Cement Concr. Res., 7, 9 - 18 (1977).

[7] L. Struble, J. Skalny and S Mindness, "A Review of the Cement-Aggregate Bond", Cement and Concr. Res. **10**, 277 - 286 (1980).

[8] D. B. Barnes, S. Diamond and W. L. Dolch. "Micromorphology of the Interfacial Zone Around Aggregates in Portland Cement Mortar", Jour. Amer. Ceram. Soc. **62**, 21 - 24 (1979).

[9] J. Grandet and J. P. Ollivier. "Nouvelle Méthode D'étude des Interface Cement-Granulats", 7th Intl. Congress on the Chemistry of Cement (Paris) Vol. III, pp VII 85 - VII 89 (1980)

[10] J. Grandet and J. P. Ollivier. "Orientation des Hydrates au Contact Granulats", 7th Intl. Congress on the Chemistry of Cement (Paris) Vol III, pp V11 85 - VII 89 (1980).

[11] K. L. Scrivener and P. L. Pratt, "A Preliminary Study of the Microstructure of the Cement/Sand Bond in Mortars". Proc 8th Intl. Congress on the Chemistry of Cement, Vol. 3, 466 - 471 (1986).

[12] C. Carde, R. Fransoise and J. P. Ollivier (1997). "Microstructural Changes and Mechanical Effects Due to the Leaching of Calcium Hydroxide from Cement Paste in "Mechanisms of Chemical Degradation of Cement-Based Systems" ed., K. L. Scrivener and J. F. Young. E and F. Spon, London. ISBN 0 419 21570 0.

[13] A. Delagrave, B. Gerard and J. Marchand (1997). "Modelling the Calcium Leaching Mechanisms in Hydrated Cement Pastes".

[14] J. R. Johnstone and F. P. Glasser. "Carbonation of Portlandite Single Crystals and Portlandite in Cement Paste". Proc 9th Intl. Cong. On the Chemistry of Cements (New Delhi) Vol 1, 125 - 154, (1992).

[15] M. Honda and S. Shizawa. "Influence of Carbonation on the Mechanical Properties and Texture of Hardened Mortar in Cement Technology" (Ed. E. M. Gartner and H. Vchikawa) pp 203 - 212 (1994). Ceramic Trans. **40** American Ceramic Soc., Westerville OH.

[16] F. P. Glasser, *et al.* "The Chemistry of Blended Cements and Backfulls Intended for Use in Radioactive Waste Disposal. Environment Agency R. and D. Report P98 (1999). ISBN 1 1857 05 157 2. Environment Agency, Swindon, England.

Table 1: Determination of $Ca(OH)_2$ in hydrated paste

Method	Brief Description and Comment
Chemical Extraction (Franke)	$Ca(OH)_2$ and CaO, if any, are selectively dissolved in an organic solvent miscible with water and titrated with standard acid. Dissolution may attack other phases, albeit slowly.
Thermal analysis and/or calorimetry	ΔH of decomposition and/or mass change measured for $Ca(OH)_2 \rightarrow CaO + H_2O$. Other reactions may contribute to signal, giving uncertain baselines. Other double salt paste constituents may decompose upon heating, contributing $Ca(OH)_2$.
Powder X-ray methods, including Rietveldt	Quantitative X-ray diffraction difficult because of the presence of mixtures of crystalline and amorphous phases, e.g., C-S-H. Fine-grained $Ca(OH)_2$ may give broadened reflections which, in conjunction with base line uncertainties, has potential for error.

Table 2. Calcium-rich limits of C-S-H

Method of Formulation	Maximum Ca/Si Molar ratio	Additional Remarks
Laboratory, by reaction of $Ca(OH)_2$ and silica gel or by coprecipitation.	1.7 to 1.9, with mean value to 1.8 □ 0.1. Once this ratio is exceeded, $Ca(OH)_2$ appears. $Ca(OH)_2$ is abundant at C/S = 2.0.	Detection of a slight excess of $Ca(OH)_2$ in the presence of an amorphous (or nearly so) majority phase is difficult.
By bottle or paste hydration of commercial Portland cements or of laboratory simulants.	Ratio close to 2.0, in PC, substantially higher than for synthetic preparations.	See text: occlusion of AFM, presence of "amorphous" $Ca(OH)_2$, etc., have all been suggested to cause high ratios in PC.

Table 3. Environmentally conditioned reactions of $Ca(OH)_2$

Process	**Comment**
Dissolution	$Ca(OH)_2$ is sparingly soluble, so selective dissolution of $Ca(OH)_2$ occurs. Influence of NaC□ is slight.
Carbonation	In normal exposure, dissolution and carbonation are complimentary. $CaCO_3$ is however, less soluble than $Ca(OH)_2$.
Reaction with soluble salts	In sea water and some groundwater, Mg replaces calcium and $Ca(OH)_2$ forms gypsum and/or anhydrite.

Table 4. Factors affecting leaching of $Ca(OH)_2$ from concrete.

Factor	Comment
Physical	Tortuosity of pore network. Discontinuous pores require matrix diffusion which is slow.
Chemical	pH and composition of leachant.
Alkalis	Alkalis, effectively present as hydroxides, decrease solubility of $Ca(OH)_2$.
Magnesium	Replaces Ca forming phases of low solubility relative to $Ca(OH)_2$, e.g. $Mg(OH)_2$.
Chloride	Relatively little impact on solubility.

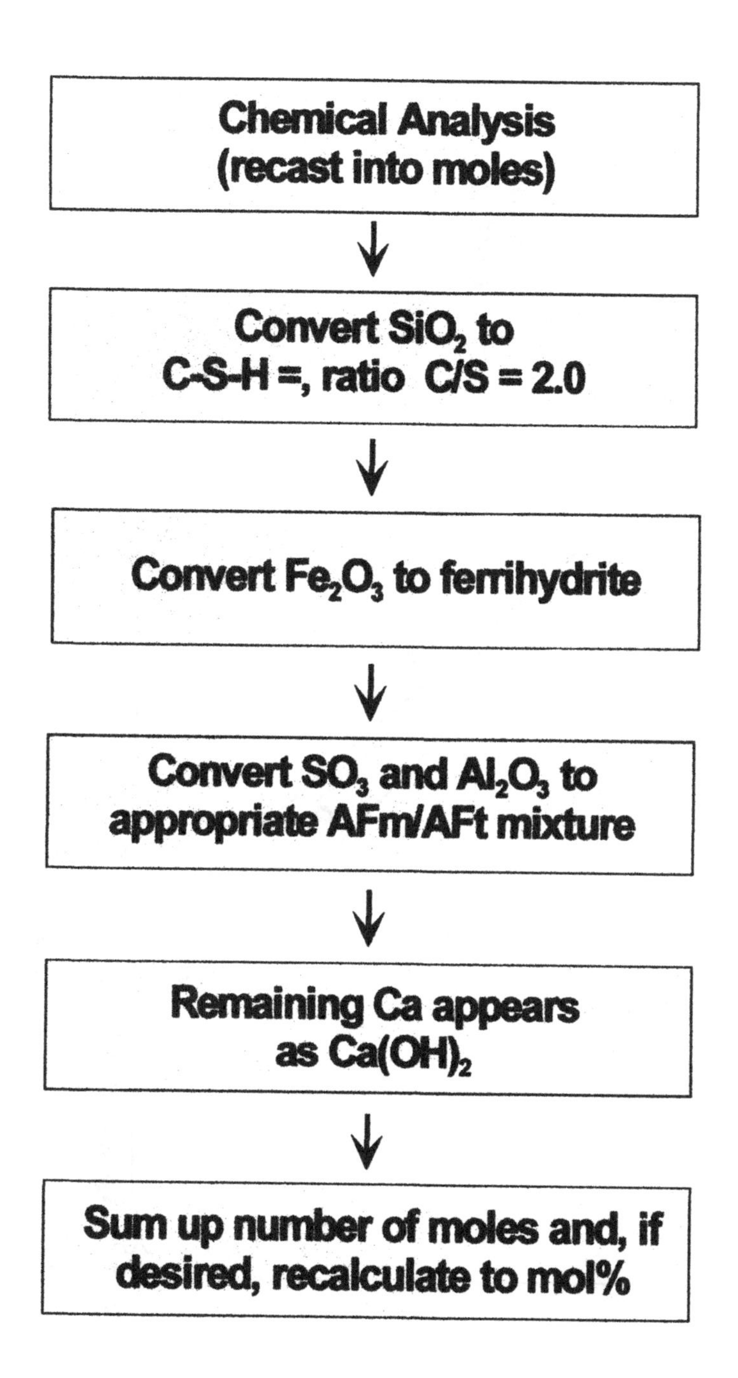

Figure 1 - Calculation of Free $Ca(OH)_2$ in Portland Cement Paste

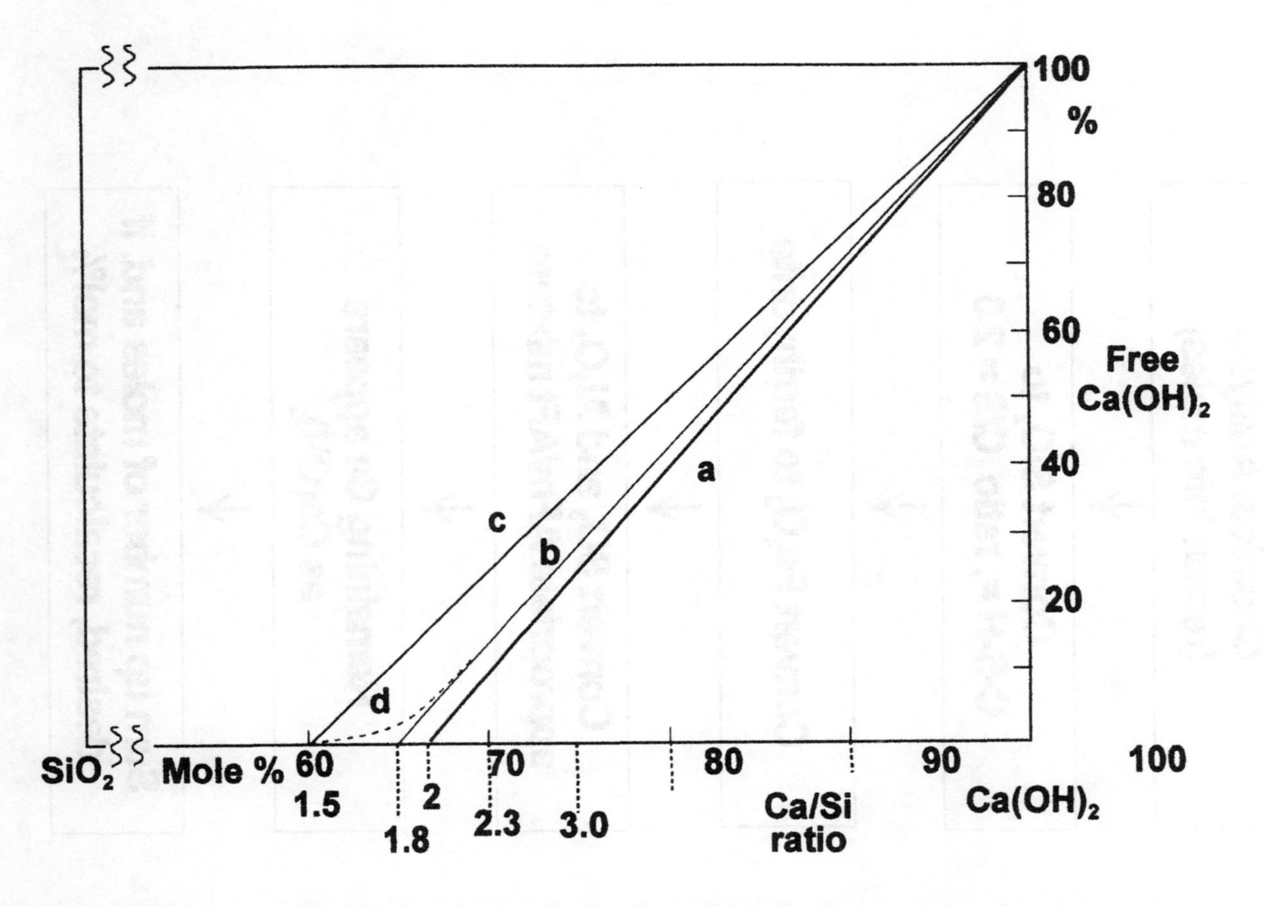

Figure 2 - Construction used to illustrate the dependence of portlandite content on cement Ca/Si ratio and Ca/Si ratio of C-S-H

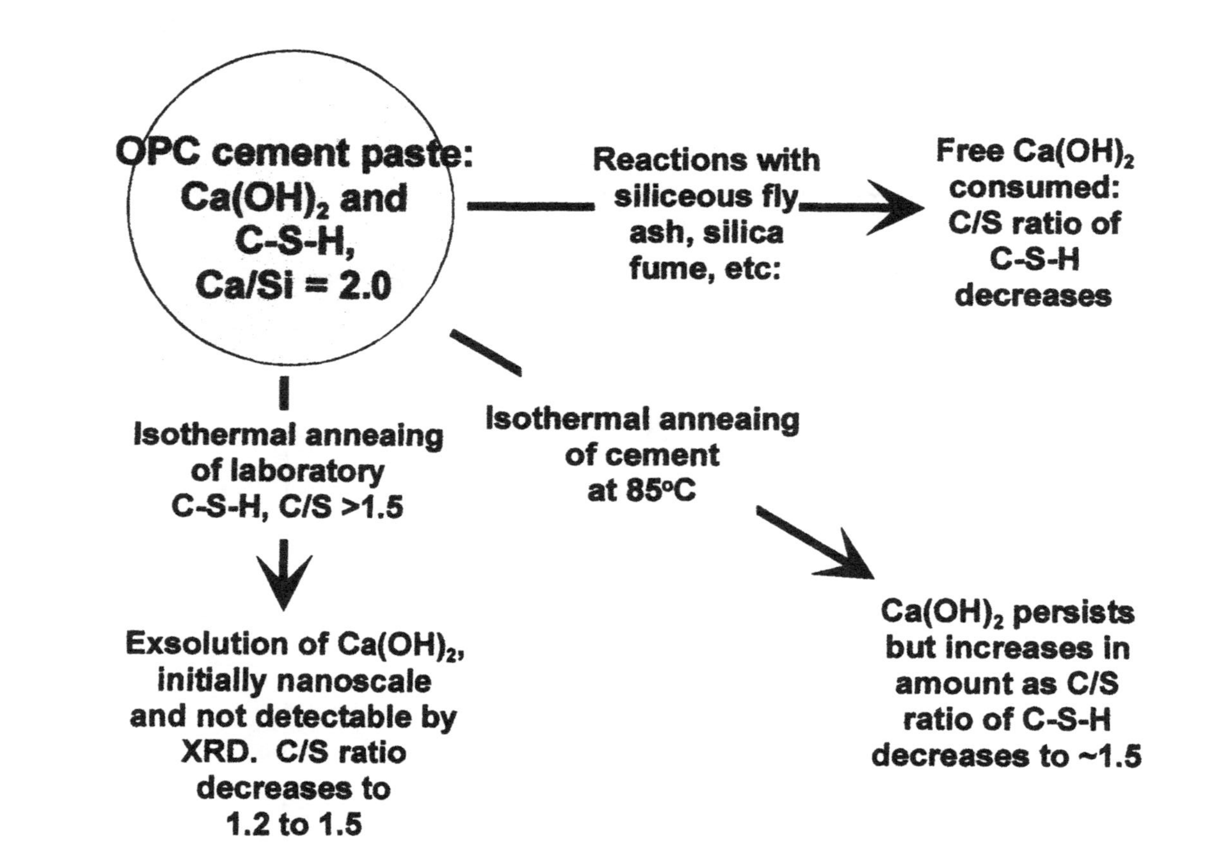
OPC cement paste: Ca(OH)2 and C-S-H, Ca/Si = 2.0
Reactions with siliceous fly ash, silica fume, etc:
Free Ca(OH)2 consumed: C/S ratio of C-S-H decreases
Isothermal anneaing of laboratory C-S-H, C/S >1.5
Isothermal anneaing of cement at 85°C
Exsolution of Ca(OH)2, initially nanoscale and not detectable by XRD. C/S ratio decreases to 1.2 to 1.5
Ca(OH)2 persists but increases in amount as C/S ratio of C-S-H decreases to ~1.5

Figure 3 - Schematic, showing pathways whereby high ratio C-S-H contributes labile portlandite.

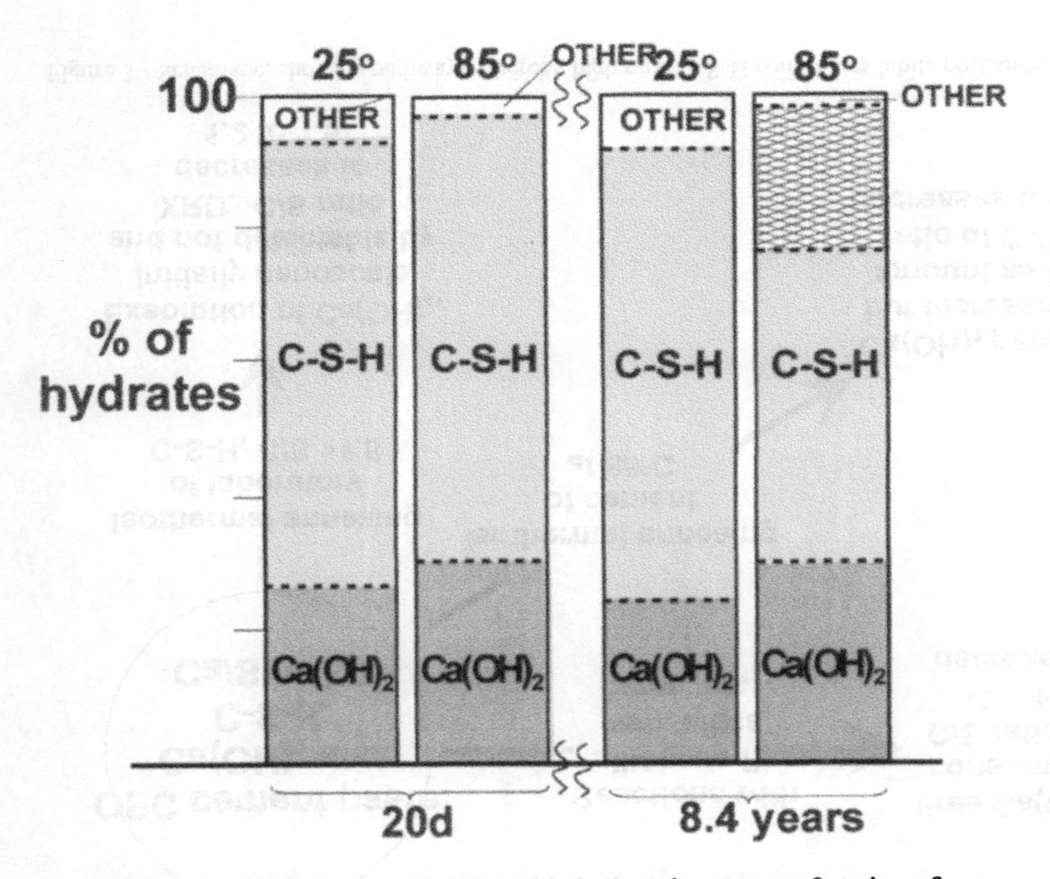

Figure 4 - Mineral balances achieved in hydrated paste as a function of cure duration and temperature.

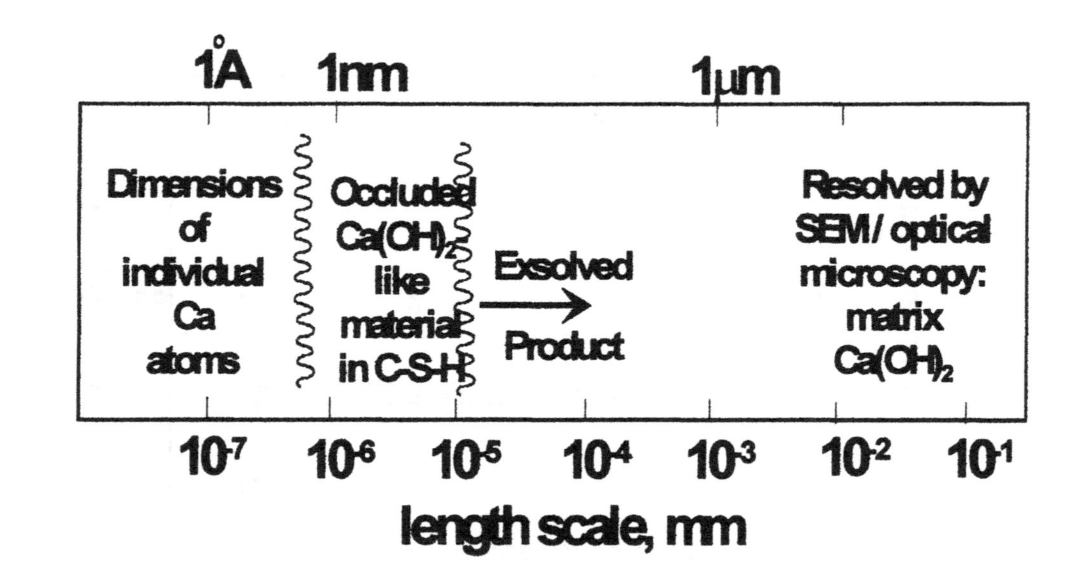

Figure 5 – Schematic, showing length scales of $Ca(OH)_2$ in hydrated cement paste

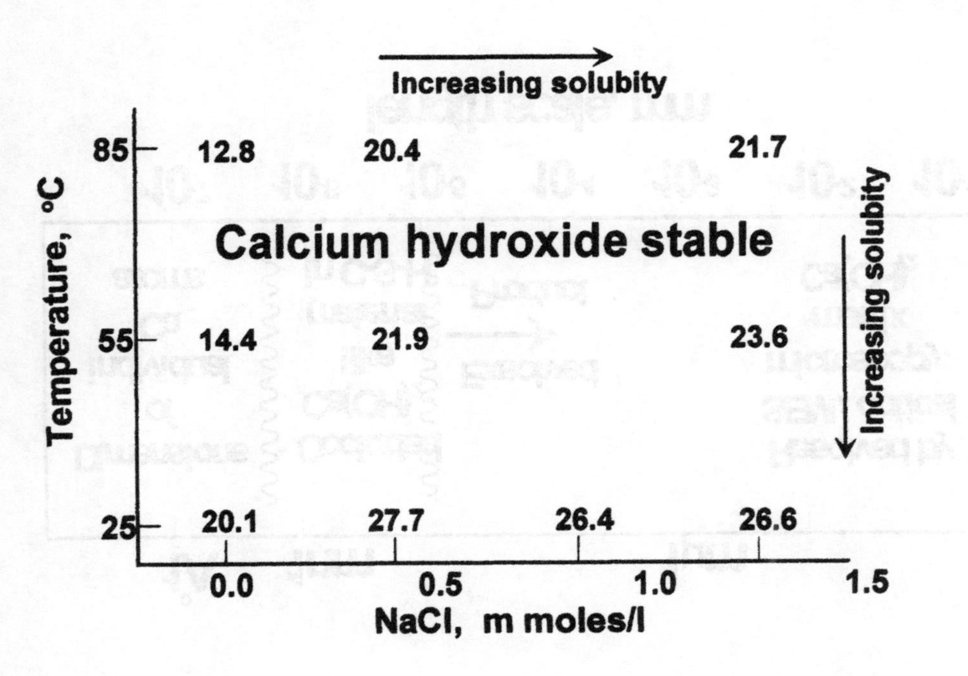

Figure 6 - Solubility of $Ca(OH)_2$ in NaCl and water as a function of temperature and NaCl concentration. Sea water is approximately 0.45M in NaCl.

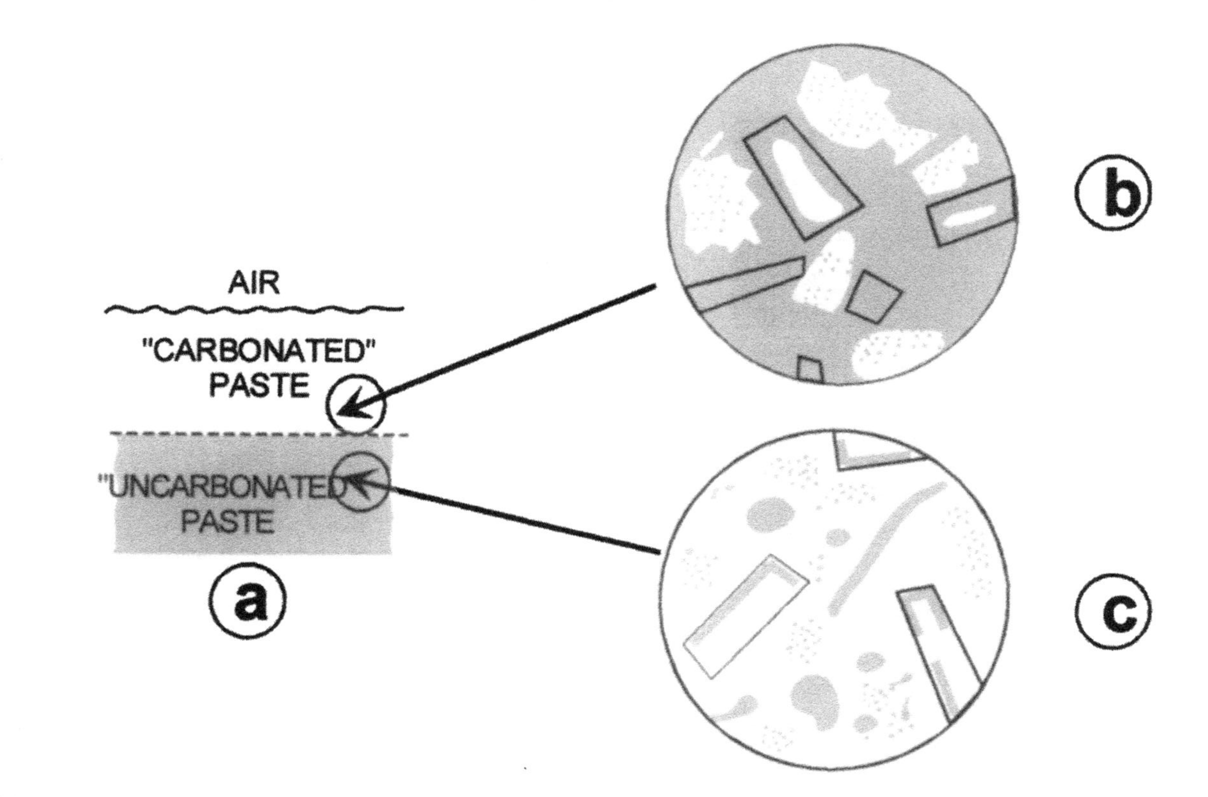

Figure 7 - Progress of carbonation in a cement paste showing (a) the apparently sharp front revealed macroscopically by the pheolpthalein test and, b and c, at higher magnification. The zones adjacent to the front. Portlandite crystals or former portlandite crystals are shown in the matrix of a highly carbonated paste, b. Some C-S-H, mainly inner hydrate, escapes carbonation as well as cores of larger and more isolated portlandite. At depth, below the pheolpthalein front, portlandite may be partially carbonated. Carbonation also affects the matrix particularly along microcracks and interconnected porosity.

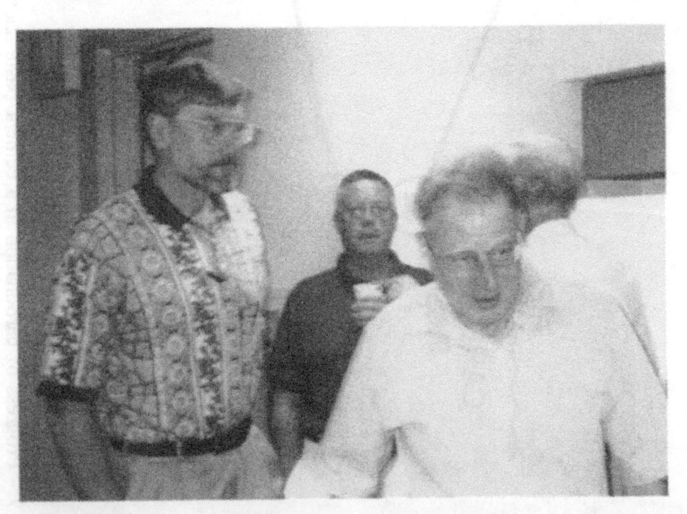

Jacques Marchand, Jim Beaudoin, Juri Gebauer (in background), Jan Skalny, and Greg Barger

CALCIUM HYDROXIDE IN CEMENT PASTE AND CONCRETE - A MICROSTRUCTURAL APPRAISAL

Sidney Diamond
School of Civil Engineering
Purdue University
West Lafayette, IN 47907-1284

ABSTRACT

The microstructural role of calcium hydroxide (CH) in cement paste and concrete is reviewed. A number of backscatter SEM micrographs are used to provide illustrations of typical appearances and configurations of CH particles in hydrated cement paste. Image analysis assessments of CH contents in cement paste are illustrated, and the results are shown to be comparable to CH contents of similar pastes determined by DTA. The results of quantitative measurements of the size distributions of CH particles in cement paste, as determined by image analysis, are provided. Determinations of several particle shape parameters for CH particles in cement paste are illustrated. Differences between CH in cement paste and CH in concrete are discussed. The presence of CH deposits on some of the surfaces of sand and coarse aggregate grains and around air voids are illustrated. The vulnerability of the CH deposits on aggregate surfaces to dissolution in permeable concretes is discussed, and possible effects of such dissolution noted.

INTRODUCTION

Calcium hydroxide plays various chemical and microstructural roles in cement paste and in concrete. While the chemistry of calcium hydroxide and its influences on cement behavior are well understood, the microstructural aspects of CH in cement paste and more particularly in concrete, are not widely appreciated. The availability of backscatter-mode scanning electron microscopy provides a tool for exploring the morphological features of this phase, and image analysis methods provide tools for assessing certain of its features quantitatively. In the present paper the writer has attempted to illustrate the morphological characteristics of CH in cement paste and in concrete, and to provide some indication of their significance. The scale of the features illustrated is that appropriate for the examination of most particles in hydrated cement paste, from about half a μm to ca. 100 μm.

CALCIUM HYDROXIDE IN PLAIN CEMENT PASTE: MORPHOLOGY

Calcium hydroxide is an unusual component of cement paste in that, unlike other cement hydration products, it is well crystallized and gives rise to strong well-defined x-ray diffraction patterns. It is also unusual in that it appears to be always stoichiometric in composition, and is not subject to solid solution effects. These features have sometimes misled workers into supposing that calcium hydroxide in cement paste is composed of 'good' crystals, i.e. crystals delineated by hard-edged boundaries, if not actually euhedral in outline. Examination of the appearance of calcium hydroxide in hydrated cement systems by backscatter-mode SEM indicates clearly that this is not the case.

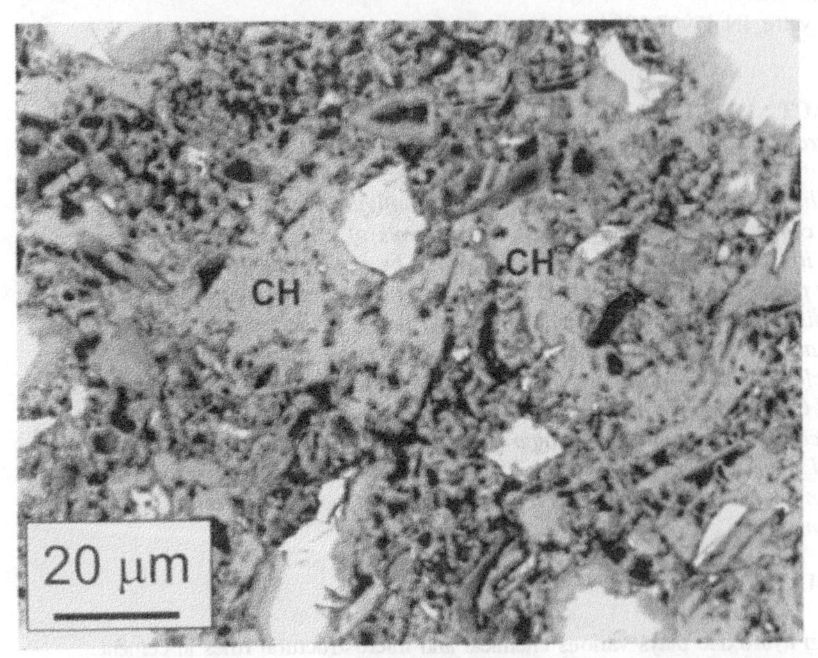

Fig. 1. Backscatter SEM image of an area in a 7-day old w:c 0.45 cement paste, illustrating appearance of various CH particles.

Fig. 1 illustrates the appearance of CH in an unusually porous area of a 7-day old w:c 0.45 plain portland cement paste. The dark areas in the figure are epoxy resin-filled pores; the very bright features are residual unhydrated cement. The intermediate gray level areas are of two kinds - slightly darker C-S-H of several morphologies, and slightly brighter CH. The distinction between C-S-H and CH is consistent, if not always easy to make.

Some of the CH deposits, so marked in the figure, are clustered in a band across the center of the micrograph. The boundaries of these masses do not resemble the hard-edged, mostly straight-line boundaries of the unhydrated cement grains in the same figure. Rather, the CH crystals show complicated, tortuous, and in some places indefinite boundaries at the scale of the examination. In Fig. 1 there is an interrupted band of CH particles running across the middle of the field which range in size from ca. 5 μm to ca. 25 μm. It is difficult to be certain of whether all of these are separate CH particles or whether some of them are part of a single large mass of CH joined together at a level above or below the plane of the image.

Fig. 2, taken from the same cement paste, shows an area much more sparsely occupied by CH. The CH particles here are fewer and smaller, and they have more definite and less convoluted perimeters than does most of the CH in Fig. 1. Nevertheless, they have unusual shapes. For example, attention is drawn to

Fig. 2. Backscatter SEM image of a different area of the same paste shown in Fig.1. The CH particles here are smaller and better defined, but still of complex shapes.

the elongated CH particle in the middle area near the bottom of the figure. This particular particle is fairly long (ca. 20 μm long), but narrow, and "pinched" in the middle. It almost certainly represents a crystallization of CH into a narrow space between the bright, partly-hydrated cement grain immediately above it and two similar, but fully-hydrated cement grains below it. The boundaries of the CH particle seem to have conformed to the available space between the original cement grains.

Some further insight into how CH particle morphologies develop may be gained from examination of Fig. 3, taken from a 1-day old w:c 0.45 cement paste. At this early stage of hydration the pore space is prominent. Much of it represents space within small thin-walled hollow shells that are linked together. These shells are remnants of small cement particles that have undergone rapid and complete hollow-shell hydration. The structure corresponds to what was described as 'reticulated network' or 'Type II' C-S-H for many years. The CH particles appear to be superimposed on this network structure, but obviously the CH is growing through rather than on the C-S-H structure.

Attention is called to the quite distinct CH particle in the lower left corner of the field in Fig. 3. The immediate area of this particle is shown at higher magnification in Fig. 4. Close examination suggests the perimeter of this CH crystal tends to closely follows the boundary walls of the linked C-S-H network through which it has grown.

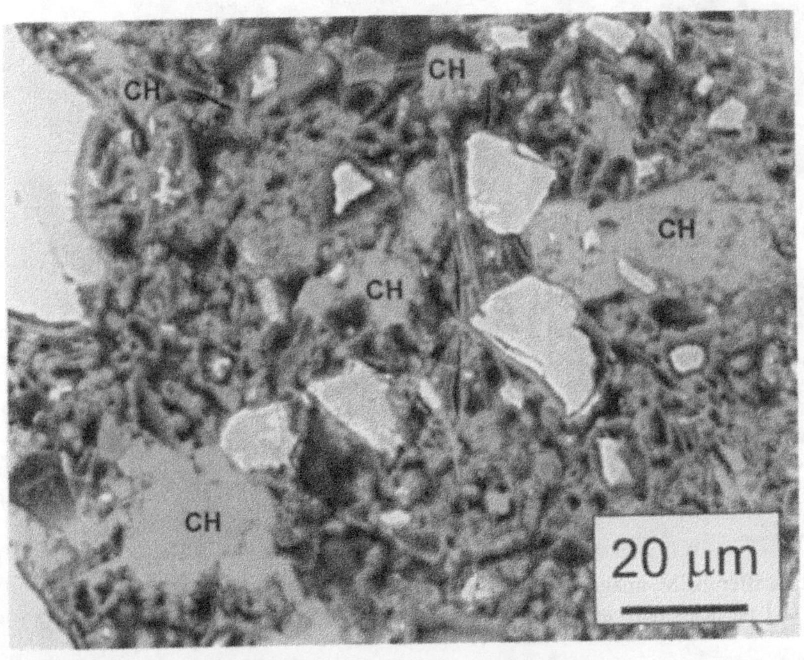

Fig. 3. Backscatter SEM from a 1 day old w:c 0.45 cement paste showing CH deposited in an area of linked thin-walled C-S-H shells.

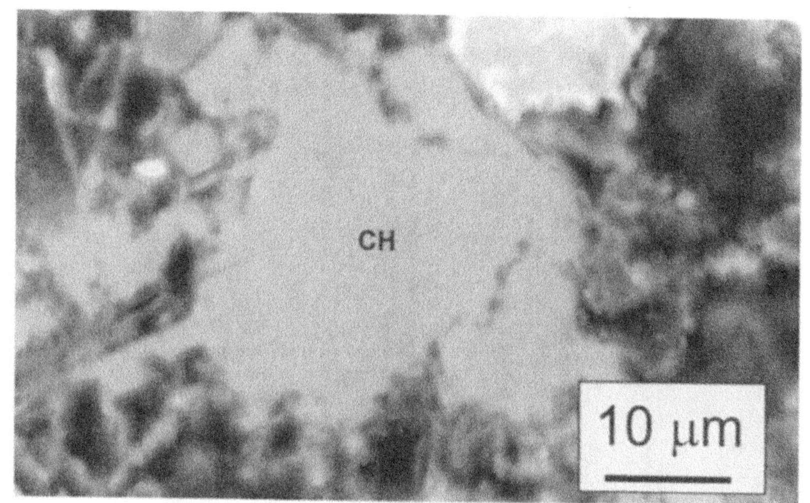

Fig. 4. Enlarged view of portion of Fig. 3.

The irregularity and geometrical complexity of the boundaries of the CH particles in young cement pastes appear to be maintained at later ages; further hydration produces a denser overall structure, but the CH within it does not appear to recrystallize extensively or show changed features. Fig. 5 shows a representative region in a 100-day old w:c 0.45 cement paste at lower magnification, with a few of the CH particles marked. Neither the sizes nor the perimeter complexities of the CH differ here from those of the young cement pastes.

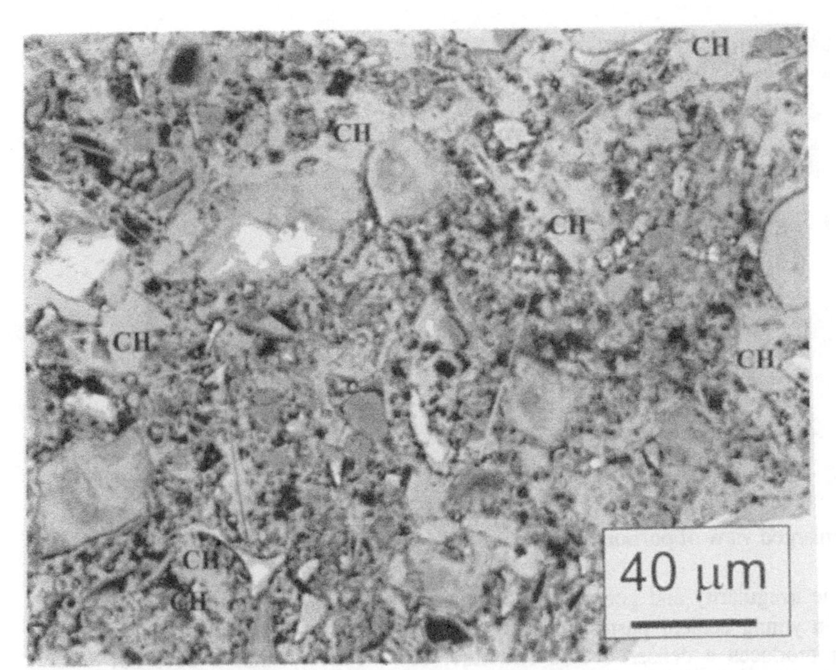

Fig. 5. Backscatter SEM image of a field in a 100-day old w:c 0.45 cement paste, showing that the characteristic appearances of CH are retained at later ages.

Attention should be drawn to one apparent discrepancy between the present observations and those appearing in secondary mode SEM micrographs taken many years ago. All of the images shown here are taken on randomly cut and polished surfaces, and should be representative of the typical range of microstructural features. In earlier years of SEM examinations of cement paste, images could be made only on surfaces of fractured specimens, using secondary-mode SEM. While clear images at relatively high magnifications were readily obtained, the representativeness of what was being imaged was often questioned. In papers describing such examinations, for example [1], especially in young pastes, CH appeared often as thin (ca. 1 μm) layered crystals, with at least some euhedral surfaces. The small and more nearly euhedral crystals depicted were quite different in appearance from the much larger irregular CH bodies illustrated in the present paper. Undoubtedly euhedral CH particles do exist in cement pastes, but it now seems likely that they are not very common.

EFFECT OF SUPERPLASTICIZER ON THE MORPHOLOGY OF CALCIUM HYDROXIDE

As indicated in a previous paper [2] the incorporation of superplasticizer in cement paste has a profound influence on the morphology of CH in cement paste.

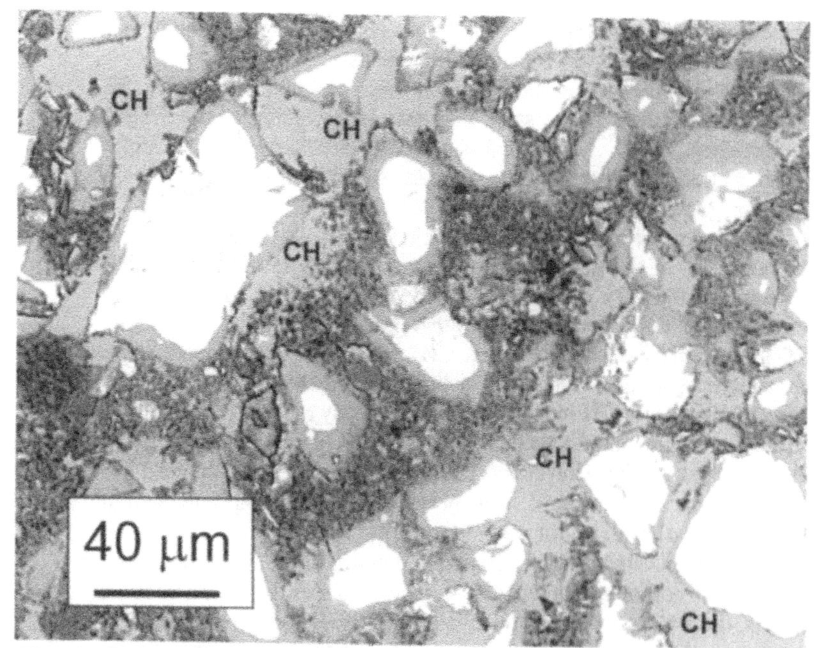

Fig. 6. Backscatter SEM image from a 100-day old w:c 0.45 cement paste incorporating 1% Mighty™ superplasticizer, showing the strong effect on CH morphology.

Fig. 6 shows a representative area of a 100-day old w:c 0.45 paste incorporating 1% of Mighty 100™ naphthalene sulfonate superplasticizer at the same magnification as Fig. 5. A comparison with Fig. 5 makes evident the strong effect of the superplasticizer on CH particle sizes and morphologies. In Fig. 6 large dense masses of CH (ca 20 - 40 µm) are seen to entirely fill substantial spaces between widely separated cement grains. In contrast to much of the CH in the plain cement pastes depicted earlier, most of these CH particles are sharp-edged, and they are much less convoluted.

QUANTITATIVE IMAGE ANALYSES OF CH IN CEMENT PASTES: ASSESSMENT OF OVERALL CH CONTENT

Image analysis of backscattered SEM images of cement paste and concretes can be used to provide quantitative descriptions of the content and characteristics of the various components in cement-based systems. The techniques, pioneered by Scrivener et al. [3], have been applied to a variety of problems. In such analyses, the gray-scale images acquired in backscatter SEM examinations undergo binary segmentation by gray level. Individual pixels constituting the image are commonly segmented into pores, unhydrated cement grains, and CH. Residual pixels not falling into one of these categories are usually assigned as 'C-S-H', although the C-S-H so delineated may contain very fine pores, ettringite, and various other minor components.

In such analyses the percentage of each kind of pixel may simply be calculated, or connected pixels of the same kind may be recognized as individual features (i.e. particles). Sizes, shapes, and other characteristics of the particles may then be determined. The size and shape measurements are necessarily those exhibited on the image plane in two dimensions.

The reliability of the results depends on a number of factors, including especially the adequacy of the gray level segmentation. Gray level segmentation poses no problem for the very bright unhydrated cement, and is comparatively straightforward for the dark pores, at the other end of the gray level scale. However, the gray level distinction between CH and C-S-H is somewhat more difficult, the calcium hydroxide being only a little bit brighter than the C-S-H. Accordingly, segmentation for CH may be less than perfectly reliable.

Nevertheless, image analysis techniques have been applied to the study of CH in cement pastes at Purdue University, and a few of the results obtained are summarized here. Some of them have been previously published in conjunction with quantitative results of analyses for other phases [4].

The least complicated image analysis determination is simply the measurement of the overall area percentage of a given phase. In the absence of perturbing effects, the area percent is an adequate estimate of volume percent. The volume percent can then be converted, with appropriate assumptions, to an estimate of mass percent. In expressing mass percentages of components in hydrated cement systems it is usually convenient to express them on an ignited weight basis, equivalent to the mass percent with respect to the original cement. Otherwise the varying amounts of non-evaporable water incorporated in the base on which the percentage is calculated unduly complicate comparisons.

According to Taylor [5], for portland cement pastes cured for several months, the expected range of CH contents, expressed on an ignited weight basis, is 15% - 25%.

Calculating mass percent values from area percent values is a convenient method to compare the results of image analysis determinations with determinations made by more established methods. The only parameters required to do the calculations are the densities (in g/cm^3) of the cement (taken as 3.15), of water, and of CH (taken as 2.44). It is assumed that the modest volume change due to paste shrinkage on drying, and the modest content of entrapped air in the paste can be ignored. It is convenient to select a base value of the mass of cement for calculation purposes, as for example 100 g. The volume of cement paste corresponding to 100 g. of cement is easily calculated from the w:c ratio and the known densities of cement and water. The absolute volume of CH in this unit of cement paste is the calculated paste volume multiplied by the volume percent of CH, as determined from the image analysis procedure. The corresponding mass of CH (per 100 g. cement) is simply the volume of CH multiplied by its specific gravity, 2.44.

Image analysis determinations were carried out for the CH contents of cement pastes made from an ordinary Type I cement at w:c ratios of 0.30 and

0.45, and hydrated for 3 days and for 100 days. The 100-day results have been published previously [4].

Table 1 provides the area % results, and their equivalents in mass percent on an ignited weight basis. The CH area percent value are on the order of 10% to 13%; the corresponding calculated mass percent CH values are between 17% and 21%, i.e. well within the expected range suggested by Taylor [5]. Table 1 also contains

Table I. Calculations of Mass % CH (Ignited Weight Basis) from Image Analysis Determinations

w:c ratio	Age, days	Area % by Image Analysis	Equivalent Mass %	Mass % By DTA
0.30	3	12.5	17.3	---
0.30	100	13.2	18.2	---
0.45	3	10.8	18.1	---
0.45	100	10.1	20.8	---
0.49	3	---	---	15.7
0.49	90	---	---	20.9

results of some previously-published determinations of CH in cement pastes by DTA [6], which provide some further indirect indication that the image analysis results are at least approximately correct. These DTA determinations were carried out some years before the image analysis experiments. They were made on w:c 0.49 cement pastes made from a cement from the same cement source and bearing almost the same chemical analysis as the cement used later for the image analysis samples. The DTA results, expressed on an ignited weight basis, are within a percent or two of the image analysis results for the 3-day old pastes; the 90-day DTA values are essentially identical to the image analysis values for 100 days. Thus, despite the uncertainty associated with binary segmentation of CH, the image analysis procedure does appear to provide accurate values for CH content.

IMAGE ANALYSES OF CH IN HYDRATED CEMENT PASTES: ASSESSMENT OF SIZES AND SHAPE CHARACTERISTICS

It is apparent from the SEM images presented previously that the CH masses found in plain portland cement paste are variable in size, have complicated shapes, and in some cases fairly indefinite outlines. Nevertheless, it is possible to apply standard methods of feature analysis to provide some attempt at quantitative evaluation of the size and shape characteristics of CH particles in cement pastes.

Particle Sizes and Particle Size Distributions

Even the definition of "particle size" is complicated for CH in cement pastes. First, the difficulty of fixing the boundary of some of the CH particles has already been alluded to. Second, the size recorded is necessarily the size detected on the two-dimensional image. Third, the difficulties of defining what is meant by particle size itself are compounded when the particles have complex shapes, as most of the CH particles in cement paste do. Thus, a clear operational definition of what is meant by 'size' needs to be adopted.

In conformity with the usual practice in image analysis, the size of a particle in this work is taken as the mean of the lengths of twelve measured 'diameter' lines running across the centroid of the particle on the image plane, the lines being taken at successive 15° rotations. Fig. 7 provides an illustration.

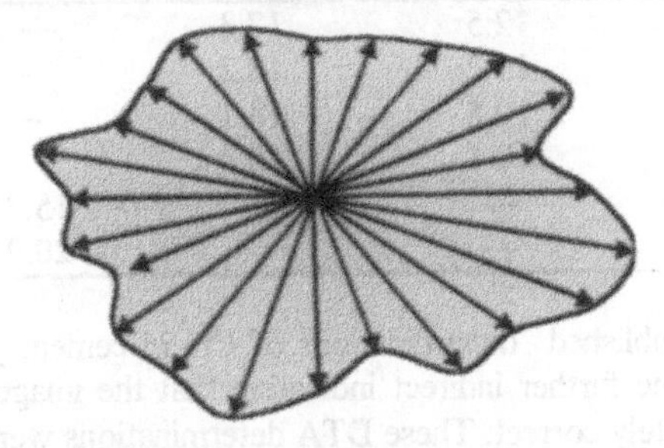

Fig. 7. Illustration of the procedure used to determine the "size" of a particle of irregular shape.

For highly elongated particles or particles of complex shapes this definition of size does certainly not convey a perfect representation, but it appears to be the most unbiased of the measurements that might be made.

Several thousand CH particles in several different cement pastes were "sized" by this procedure, primarily to generate data for particle size distributions. Size distributions compiled this way are number-based distributions, in which each particle is accorded the same importance regardless of its size. In most number-based size distributions, fine particles become particularly important, since there are usually so many of them. However, in image analyses, the sizes of the finest particles that can be detected, i.e. those only a few pixels in size, are of dubious validity; they are often arbitrarily removed from the image by preliminary image processing steps. We chose not to do this, but instead limited the minimum diameter of the features tallied in the size measurements to 4 μm.

Fig. 8 provides plots of the size distributions for recognized CH particles in w:c 0.45 cement pastes at ages of 3 and 100 days. Note that in these distributions only a few particles are tallied in the 4 to 5 μm class, compared to those in 5 to 6 μm class. This suggests that, dropping the suspect particles below 4 μm did not bias the results unduly, assuming that the size distributions have the normal shapes .

As can be seen in Fig. 8, there is not much difference between the CH size distributions for the two ages; the size distribution for the older paste is actually slightly finer. The mean size values for the number distributions are 7.7 μm for the 3-day old paste and 6.7 μm for the 100 day-old paste. Both distributions peak sharply between 5 and 6 μm and tail off to larger sizes. Occasional CH particles of diameters up to about 15 μm are found in both, and isolated 'outliers' up to about 25 μm were occasionally tallied.

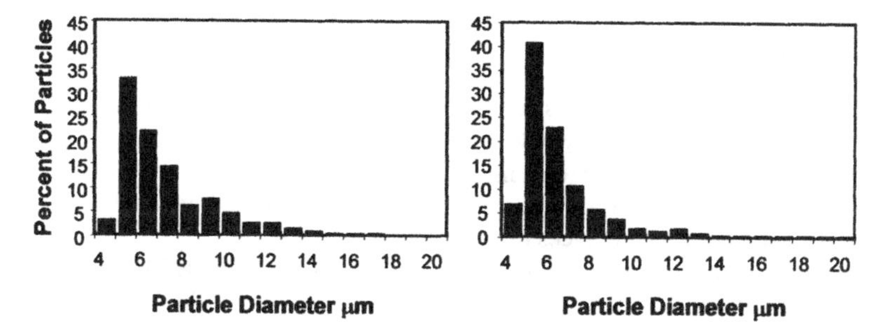

Fig. 8. Size distribution histograms for CH particle size distributions in 3- and 100-day old w:c 0.45 cement pastes.

Data are also available for corresponding cement pastes of w:c 0.30. Their size distributions are similar to those of the w:c 0.45 pastes shown in Fig. 8, but are slightly narrower, i.e. they show fewer of the occasional large particles. The respective distribution mean size values are 6.1 μm and 5.8 μm for 3-day and 100-day w:c 0.30 pastes.

Shape Measurement of CH Particles: Elongation and Degree of Convolution

Standard image analysis techniques provide various methods of quantitatively assessing certain aspects of particle shapes. In view of the complex shapes of CH particles seen in the micrographs presented earlier, it was considered appropriate to apply some of these methods to the CH particles in the cement pastes studied.

One shape measurement thought to be of interest was that of particle elongation, since many of the CH particles depicted earlier were visibly elongated. The measurement used, rather confusingly called "circularity", is a parameter given by the product of pi times the square of the maximum diameter of the particle, divided by 4 times its measured area. This parameter has a value of 1.00 for a circle; progressively more elongated particles have increasingly higher values. For example, an ellipse has an elongation parameter value that is simply the ratio of its major to minor semi-axes. Thus for an ellipse with a 3:1 ratio of semi-axis lengths the circularity is 3.

For the populations of CH particles in our various cement pastes, the mean values of circularity were all similar to each other and were between 3.2 and 3.6.

This indicates that fairly elongated particles are common in all the cement pastes. However, there is wide scatter in the values measured for individual CH particles within any specific cement paste. The general range of the circularities of CH particles within a given paste was between values about 1.5 (indicating little elongation) and as much as 6 or 7 (indicative of highly elongated particles).

In assessing the significance of such values, it should be recalled that they reflect elongations measured on an arbitrary plane surface; their counterparts in three dimensional space are not readily determinable.

Another shape parameter that was thought to be of interest is a standard image analysis measurement of the degree of convolution of particle perimeters called "form factor". This measurement is likely to provide realistic values representing three-dimensional reality, since a particle showing a highly convoluted boundary in two dimensions would reasonably be expected to show a similar degree of convolution in three dimensions. Many of the CH particles depicted in the previous illustrative of cement paste morphology had visibly complex shapes and apparently high degrees of convolution.

The form factor of a particle is defined as 4 pi times its area, divided by the square of its measured perimeter. A circle has a form factor value of 1.00, but for firm factor measurement, departures from circularity reduce, rather than increase the value. Elongated but not convoluted shapes show modest reductions from 1; for example, an ellipse of semi-axis ratio 2:1 has a form factor of 0.8; one of semi-axis ratio of 3 has a form factor of 0.6. The degree of convolution of the perimeter has a much greater effect on the measurement, and the form factor decreases sharply as the length of the perimeter increases with the extent of convolution.

Analyses were carried out automatically for all of the CH particles identified in both w:c 0.45 and w:c 0.30 cement pastes at ages of 3 and 100 days, numbering several thousands of particles in all. The mean form factor values for all four pastes were similar to each other, all being between 0.44 and 0.46. It appears that the “average” degree of convolution of CH particles in cement paste does not vary significantly with age or w:c ratio.

However, as might be expected from observations of the images presented earlier, the measured form factor varies with the size of the CH particle. It is smaller, on average, for larger particles, i.e. the extent of convolution of the perimeters of large particles is greater. However, averages can be misleading. Fig. 9 shows a scatter plot of form factor vs. particle diameter for the CH particles found in the 100-day old w:c 0.45 paste.

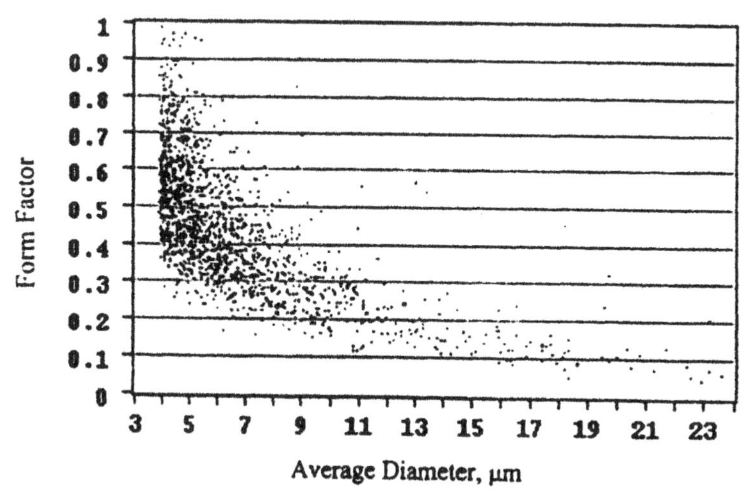

Fig. 9. Scatter plot of form factor vs. particle diameter (average of 12 measurements) for CH particles in 100-day old w:c 0.45 cement paste. Each dot represents 1 particle.

Smaller particles, e.g. those of mean diameter between 4 and 10 μm, show individual form factors extending over a wide range of values, from about 0.2 to nearly 1. The interpretation is that some small CH particles are convoluted, but that many of them are not. On the other hand it can be seen that almost all of the larger particles, i.e. those with average diameter greater than about 12 μm, are highly convoluted; they exhibit form factors only ranging between about 0.1 and about 0.2. Almost all of the very largest CH particles show form factors close to 0.1, i.e. they are very highly convoluted indeed.

CALCIUM HYDROXIDE IN CONCRETES

General

Calcium hydroxide in concrete present certain features not usually found in cement paste. In particular, CH in concrete is found in close association with the surfaces of sand and coarse aggregate particles, and in association with entrained air voids, in addition to being distributed within the hydrated cement paste.

Calcium Hydroxide at Aggregate Interfaces

In assessing the characteristics of CH in concrete (as distinguished from CH in cement paste) the first major observation normally made is that CH is highly concentrated near the surfaces of coarse aggregate and sand grains, and indeed is often deposited directly on them. Fig. 10 provides an SEM backscatter mode image taken from a 3-day old w:c 0.50 concrete illustrating this feature. The sand and aggregate in this concrete were both crushed dolomite.

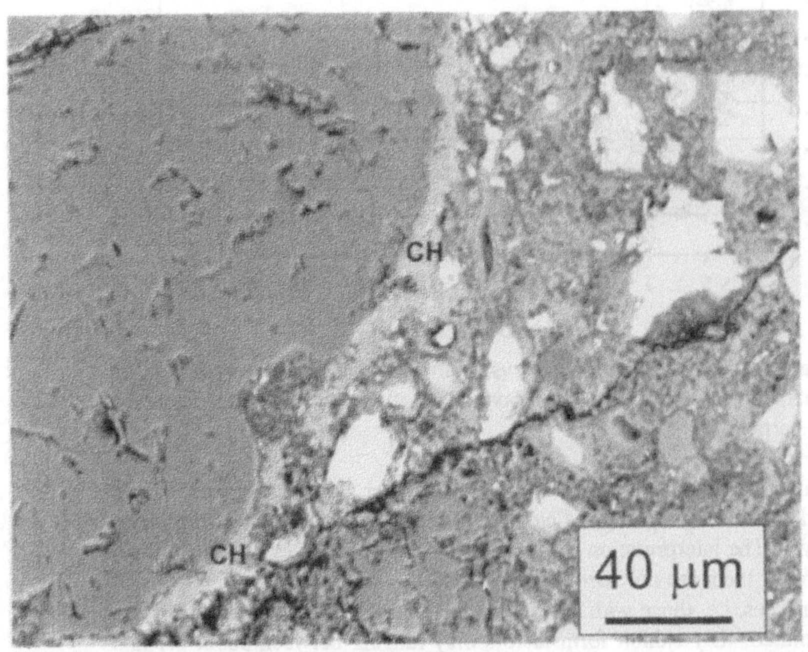

Fig. 10. Backscatter SEM image taken from a 3 day-old w:c 0.50 concrete showing CH deposited against part of the surface of a dolomite sand grain.

While there are a few small CH particles in the field away from the sand grain, the bulk of the CH depicted is in an irregular but roughly 10 µm-thick boundary layer in direct contact with the perimeter of the aggregate. The inner boundary of the CH layer is mostly smooth and follows the contact with the aggregate surface quite accurately. The outer boundary, toward the paste, is irregular and convoluted. The layer is interrupted roughly two-thirds of the way down the field, where an indentation occurs in the sand grain surface. The CH deposit appears to be made up of several adjacent individual CH particles, rather than being continuous.

In the writer's experience most sand grains in concretes (including field concretes of various types) have some portion of their perimeters either covered with or at least in partial contact with such deposits of CH. These deposits, sometimes extending as much as 20 µm into the cement paste, form prominent features of the interfacial transition zone (ITZ) as conventionally described. This CH layer is usually discontinuous and rarely, if ever, entirely surrounds a given aggregate. Indeed, it is common to see significant stretches of the perimeter of many aggregates entirely free of CH deposits, and occasional small sand grains may lack such deposits entirely.

In much of the early literature on the ITZ, the CH near the interface is described as preferentially oriented with respect to the local aggregate surface, with its c-axis assumed to be normal to the surface. In model specimens consisting of cement paste cast against smooth flat rock or glass surfaces the CH deposits do indeed exhibit a considerable degree of orientation, as shown many years ago by Hadley [7], and elegantly confirmed using a pole figure goniometer by Detwiler [8]. The degree of orientation is highest in the CH deposited at the contact with the aggregate, and the reduction of orientation of CH with depth into the paste was the basis of a well-known method of measuring the thickness of the ITZ developed by Grandet and Ollivier [9].

The degree of orientation of these CH layers in the usual concrete may or may not differ from that shown to exist in model systems. Whatever the case, the present writer feels that the significance attached to the degree of orientation is misplaced. In his view, the important feature of the CH layer is not whether it is oriented, but rather that, where it exists, the solid, non-porous CH serves to plug what might otherwise be open spaces nearest the aggregate in the ITZ.

It is known that CH deposits form on surfaces of a variety of mineralogical types of aggregate, including quartz, limestones, various feldspars, and others, but it is not known to what extent the aggregate mineral type might influence the details of the CH deposit. Indeed, the writer knows of no published study providing quantitative data on relative percentage of the aggregate surface covered by CH for any specific concrete. Such data would be of interest, but the problem of assessment is not trivial; in some areas the CH deposits are fragmentary or filmy, and in some areas they may be slightly displaced from, rather than in close contact with, the actual surface.

In the absence of accurate quantitative data, the writer offers the results of a brief survey carried out by examining a total of 61 micrographs covering essentially all of the perimeter length around four randomly chosen sand grains in the 3-day old w:c 0.50 concrete shown in Fig. 10. In this survey, slightly more than 50% of the total perimeter was bare; just over 25% of the total perimeter had a layer of CH judged to be less than about 5 µm thick, and slightly less than 25% of it had a CH deposit judged to be over 5 µm thick. Three of the four sand grains gave remarkably consistent percentage tallies. The fourth, a somewhat smaller sand grain than the others, had a significantly larger proportion of bare surface, and where it occurred, its CH layer was thinner. The degree to which these estimates of surface coverage might approximate the degree of coverage generally present in concrete is unknown.

CH Deposits Associated With Air Voids

It has been known for many years that CH deposits of several kinds are associated with air voids in concrete. It is common to find a very thin, oriented layer of CH lining the voids, and some of them have a CH deposit several µm thick surrounding them. In addition, in field concrete exposed to wetting and drying, entry of pore solutions into the air voids is common. Such intrusion is often marked by deposition of CH and of ettringite, separately or together. This CH, crystallizing in free space rather than in the spatially restrictive confines within cement paste, tends to be euhedral, and is thus morphologically different from other CH in concrete.

CH in 'Bulk' Paste Within Concrete

As illustrated in Fig. 10, CH exists as layered deposits on some aggregate surfaces. The question then arises as to whether the remainder of the CH found in the bulk of the cement paste in concrete is similar to the CH in plain cement pastes. The writer is unaware of any quantitative studies addressing this point, but qualitatively this appears to be so. For example, Figs. 11 and 12 show CH particles in the areas well away from the aggregate that appear to show the ranges of sizes and the complexity of shapes and extent of convolutions described earlier for cement pastes. Both figures were taken from the same concrete as Fig.10.

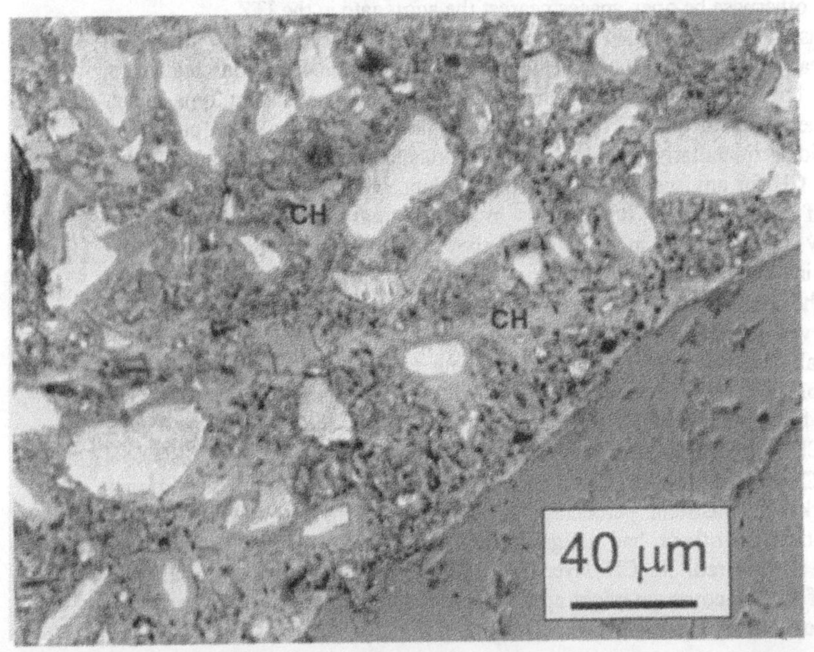

Fig. 11. Backscatter SEM image from another area of the same concrete shown in Fig. 10. This area of the aggregate has only a very thin CH deposit. CH within the paste shows elongated and convoluted outlines.

In Fig. 11 there is a only a very thin deposit of CH over the portion of the aggregate shown; in Fig. 12, there is no CH deposit on the aggregate per se; the air void almost touching it is ringed by a substantial deposit of CH.

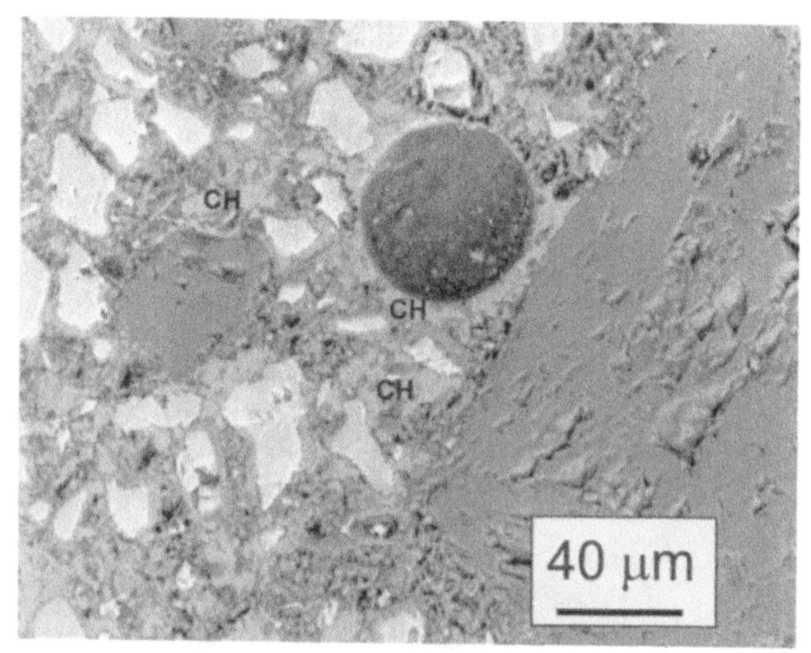

Fig. 12. Backscatter SEM image from another area of the same concrete. The aggregate surface locally is free of CH, but a deposit of CH rims the adjacent air void.

Stability of CH Deposits in Permeable Concretes

The stability of CH deposits in cases of permeable concrete exposed to external solutions is worthy of mention. The present writer and his colleagues have examined many instances of highly permeable concretes intermittently exposed to ground water solutions. In such concretes dissolution of the CH deposits adjacent to aggregates is a common occurrence [11]. CH within the bulk paste is also commonly dissolved. Fig. 13 provides a typical illustration.

Such dissolution apparently follows the progressive replacement of the original highly alkaline concrete pore solution by nearly neutral ground water. Various alteration products (including gypsum, ettringite, and others) may eventually be deposited in the vacated space. Bonen [12] suggested that preferential dissolution of CH around the aggregates is a specific feature that may open up channels for relatively easy transport of additional solution. However, the degree to which this happens may vary with different concretes, especially as a function of w:c ratio. Delagrave et al. [13] showed that when well cured mortars of w:c as high as 0.45 were subjected to uniaxial leaching, a definite front of dissolution of calcium hydroxide proceeded uniformly down the specimens, with no preferential leaching or dissolution of CH taking place around the aggregates.

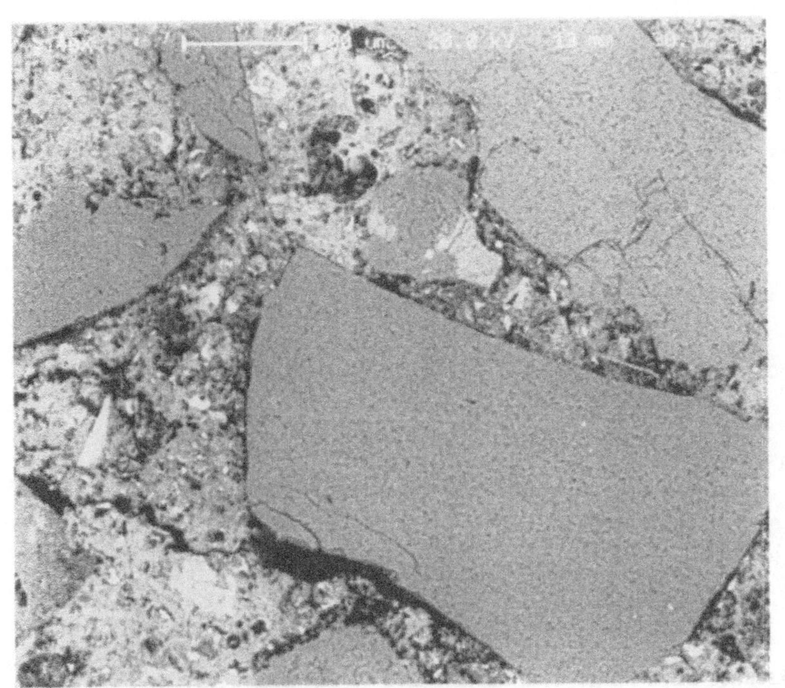

Fig. 13. Illustration of removal of CH from permeable concrete exposed to ground water for a number of years.

CH Deposits and DEF

The CH deposits normally found around aggregate surfaces are not usually found where DEF-induced expansion has created gaps around aggregates. A typical example of the appearance of such gaps in a mortar specimen that has undergone DEF is shown as Fig. 14. In such DEF specimens there are ordinarily some CH deposits around the aggregates before the DEF expansion occurs, and sometimes in the early stages of expansion ettringite and CH particles are found adjacent to each other in the developing gap. The apparent disappearance of this CH from the gaps during the later stages of the expansion suggests that the CH may have a role in the overall process. Whatever the details may be, the creation of gaps around aggregate grains in DEF as seen in Fig.14 is a quite different process from the process of dissolution of CH by penetrating ground water depicted in Fig. 13.

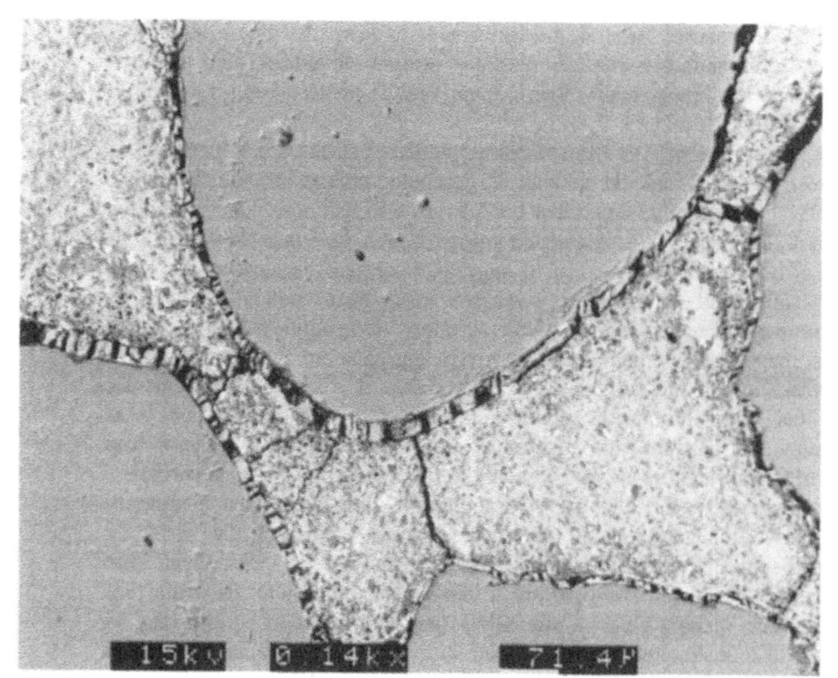

Fig. 14. Virtual absence of CH in gaps opened up around aggregates during DEF -induced expansion.

DISCUSSION

A number of specific features in the results presented appear to merit discussion.

The morphology of the CH particles found in cement paste appear to be consistent with the idea that CH deposits nucleate in whatever empty space is available to the pore solution during early hydration, and grow quickly to occupy whatever limited space is available in the vicinity. They presumably stop growing when the local space is filled. Their sizes, shapes, and convoluted particle boundaries are evidently constrained by the complex local geometry set by cement grains and early C-S-H deposits.

Special factors appear to be able to overcome these constraints. For example, the geometric constraints are largely eliminated when superplasticizer is used. This appears to be due to suppression of the normal deposition of the groundmass thin-walled C-S-H that otherwise takes place. The CH particles can then grow to occupy the full space from one cement grain to another

The relatively good agreement between the results of image analysis determinations of the CH content and the DTA determinations for similar cement pastes seem to lead to two separate conclusions. First, it appears that image analysis for CH provides satisfactory results despite the CH binary segmentation difficulty alluded to earlier. Second, the agreement suggests that there cannot be a large content of nanometer-size CH particles present, if indeed such particles exist. Such particles, being smaller than a pixel, could not have been recorded in the area % CH tally.

The appearance of CH deposits along aggregate surfaces in concrete is of major interest. It seems that the vicinity of aggregate surfaces provide a different order of relaxation of spatial constraint for CH growth than does superplasticizer incorporation. Statistically it is reasonable to infer that extra space for CH growth would tends to be available adjacent to aggregate surfaces because of the well-established wall effect. Nevertheless, such free space would not be expected to occur as continuous empty zones along the periphery of the aggregate.

It is unclear why certain areas of a given aggregate are entirely free of CH deposits while others are covered. Fig. 12 makes it difficult to accept the argument that the differences simply represent regions of differing local concentrations of calcium and hydroxyl ions, since obviously there was deposition of CH around the air void adjacent to the local bare aggregate surface.

The existence of CH deposits over areas of aggregate surfaces in concrete has been known for many years; the thicker deposits are perfectly visible in optical microscope examination, although perhaps not in the detail that backscatter SEM permits. In view of this, it remains inexplicable to the writer that even the current version [10] of the NIST 'pixel-based model' purporting to represent the ITZ in concrete and considered by their authors and many others to do so, entirely lacks representation of this important feature of concrete.

CONCLUSIONS

1. Calcium hydroxide exists in cement pastes as discrete particles ranging in 'size' from a few μm to ca. 30 μm or more, with number distribution mean sizes of the order of 7 μm. Size values can be misleading because of the extremely irregular character of many of the particles.
2. The boundaries of some of the CH particles may be "wispy" rather than hard-edged, and none of the particles within the cement pastes examined appear to be euhedral. Most particles are elongated, and the larger ones, in particular, have highly convoluted boundaries. It appears that the growth of CH nuclei is constrained and limited by coming into contact with previously-existing C-S-H or other deposits.
3. Image analysis determinations of CH content using backscatter mode SEM yield values of the appropriate magnitude for total CH, suggesting that any nanometer-scale CH present is limited in amount.
4. Concrete contains CH of morphological characteristics similar to those of plain cement paste distributed through the cement paste of the concrete, but additionally contains significant amounts of CH deposited as irregular layers of varying thickness around portions of the aggregate surface. A very crude estimate suggests that slightly less than half of the aggregate surfaces in one concrete bear such a deposit. CH deposits are also found around air voids and in field concrete CH deposits are found within air voids as well.

5. In permeable concretes in contact with ground water CH is subject to dissolution. Dissolution of the CH around aggregate surfaces in particular may leave elongated and possibly interconnected gaps or channels that could facilitate further ingress of deleterious components.
6. Gaps left by dissolution of CH in permeable concretes are different in character from ettringite-containing gaps produced around aggregates in DEF, although the latter rarely contain CH.

ACKNOWLEDGMENTS

The writer is grateful to Yuting Wang, Jingdong Huang, and Zhaozhou Zhang, former graduate students at Purdue University, who did the microstructural examinations on which this paper is based, and to Jason Weiss for technical assistance. Helpful discussions with Niels Thaulow and Jan Skalny are acknowledged with thanks.

REFERENCES

[1] S. Diamond, "Identification of Hydrated Cement Constituents Using a Scanning Electron Microscope - Energy Dispersive X-Ray Spectrometer Combination," *Cement and Concrete Research* **2** [4] 617 -632 (1972).

[2] S. Diamond and Y. Wang, "A Quantitative Image Analysis Study of the Influence of Superplasticizer on Cement Paste Microstructure," *Proceedings of the 18th International Conference on Cement Microscopy* Edited by L. Jany, A. Nisperos, and J. Bayless, ICMA, Duncanville, TX, pp.465-479 (1996).

[3] K. L. Scrivener, H. H. Patel, P. L. Pratt, and L. J. Parrott, "Analysis of Phases in Cement Paste Using Backscattered Electron Images, Methanol Adsorption, and Thermogravimetric Analysis," *Materials Research Society Symposium Proceedings v. 85*, Edited by Leslie J. Struble and Paul Brown, Materials Research Society, Pittsburgh, PA, pp. 67-76 (1987).

[4] Y. Wang and S. Diamond, "An Approach to Quantitative Image Analysis for Cement Pastes," *Materials Research Society Proceedings v. 370*, Edited by S. Diamond et al., Materials Research Society, Pittsburgh, PA, pp. 23-32 (1995).

[5] H. F. W. Taylor, "Cement Chemistry," 2nd edition, Thomas Telford, London, p. 200 (1997).

[6] S. Diamond, Q. Sheng, and J. Olek, "Evidence for Minimal Pozzolanic Reaction in a Fly Ash Cement During the Period of Major Strength Development," *Materials Research Society Proceedings v. 137*, Edited by L. R. Roberts and J. P. Skalny, Materials Research Society, Pittsburgh, PA, pp. 437-448 (1989).

[7] D. W. Hadley, "The Nature of the Paste-Aggregate Interface," Ph. D. thesis, Purdue University (1972).

[8] R. J. Detwiler, Chemical and Physical Effects of Silica Fume on the Microstructural and Mechanical Properties of Concrete, Ph. D. thesis, University of California, Berkeley (1988).

[9] J. Grandet and J. P. Ollivier, "Nouvelle methode d'etude des interfaces ciment-granulats," *Proceedings of the 7th International Congress on the Chemistry of Cement, Paris* Vol. II, pp. VII - 85 - VII-89 (1980).

[10] D. P. Bentz and E. J. Garboczi, "Computer Modelling of Interfacial Transition Zone: Microstructure and Properties," RILEM Report 20, in *Engineering and Transport Properties of the Interfacial Transition Zone in Cementitious Composites*, Edited by M. G. Alexander, G. Arliguie, G. Ballivy, A. Bentur, and J. Marchand, RILEM Publications S.A.R. L., pp. 385 (1999).

[11] S. Diamond and R. J. Lee, "Microstructural Alterations Associated With Sulfate Attack in Permeable Concretes," in Materials Science of Concrete: Sulfate Attack Mechanisms, Edited by J. Marchand and J. Skalny, American Ceramic Society, pp. 123-174 (1999).

[12] D. Bonen, "Features of the Interfacial Transition Zone and Its Role in Secondary Mineralization," in RILEM Proceedings 35, The Interfacial Transition Zone in Cementitious Composites, Edited by A. Katz, A. Bentur, M. Alexander, and G. Arliguie, E. and F.N. Spon, pp. 224-233 (1998).

[13] A. Delagrave, J. Marchand, and M. Pigeon, "Influence of the Interfacial Transition Zone on the Resistance of Mortar to Calcium Leaching,", in RILEM Proceedings 35, The Interfacial Transition Zone in Cementitious Composites, Edited by A. Katz, A. Bentur, M. Alexander, and G. Arliguie, E. and F.N. Spon, pp. 103-113(1998).

SCANNING ELECTRON MICROSCOPY IN CONCRETE PETROGHRAPHY[1]

Paul E. Stutzman
National Institute of Standards and Technology
100 Bureau Drive, Stop 8621
Gaithersburg, Maryland 20899-8621

ABSTRACT

Scanning electron microscopy has distinct advantages for characterization of concrete, cement, and aggregate microstructure, and in the interpretation of causes for concrete deterioration. Scanning electron microscope imaging facilitates identification of hardened cement paste constituents with greater contrast, and greater spatial resolution than for optical methods and provides ancillary capability for element analysis and imaging. Quantitative information may be extracted from these data, such as composition, phase abundance and distribution. Sample preparation and demonstration of some of the common imaging techniques and extraction of data using image processing and analysis are presented.

INTRODUCTION

The study of concrete microstructure has benefited greatly from the use of microscopic examination and in particular, scanning electron microscopy. Scanning electron microscopy imaging and X-ray microanalysis techniques have been developed for imaging the complex microstructure of concrete and provide images with sub-micrometer definition. Through combination of the information available in the backscattered electron and X-ray images, an accurate segmentation of the image into its constituent phases may be achieved. The application of scanning electron microscopy (SEM) enhances our ability to characterize cement and concrete microstructure, and will aid in evaluating the influence of supplementary cementing materials, evaluation of concrete durability problems, and in the prediction of service life. In this paper, the use of the

---

[1] **Contribution of the National Institute of Standards and Technology, not subject to copyright in the United States.**

scanning electron microscope will be explored through a discussion of specimen preparation and the demonstration of the commonly used imaging techniques. Their application to examination of concrete microstructure, the identification and quantitative measurement of phases will be illustrated.

The Scanning Electron Microscope

The SEM scans a focused beam of electrons across the specimen and measures any of several signals resulting from the electron beam interaction with the specimen. Images of topography can be used to study particle size, shape, surface roughness, and fracture surfaces, while polished surfaces are used for determination of phase distribution and chemical composition. X-ray microanalysis provides quantitative spot chemical analysis as well as maps of element distribution. Images are monochrome because they reflect the electron or X-ray flux resulting from the beam/specimen interaction. Backscattered electron and X-ray imaging are the most useful imaging modes for quantitative scanning electron microscopy. Computer-based image processing and analysis (IA) makes routine quantitative imaging possible.

Secondary Electron Imaging: Secondary electrons (SE) are low-energy electrons resulting from an inelastic collision of a primary beam electron with an electron of a specimen atom. Because of their low energy, they are readily absorbed and only those produced near the surface escape, resulting in an image of surface topography. The apparent shadowing in the image is a result of the absorption of the secondary electrons by intermediary parts of the specimen.

Figure 1 illustrates SE imaging of fracture surfaces of hardened cement paste. At early-ages cement pastes typically have sufficiently large void spaces that well-formed crystals can develop, while at later ages, the well-formed crystal shapes are generally found only in regions of very high porosity or in air voids. In the upper image one can see plates of calcium hydroxide, exhibiting a typical hexagonal habit, and ettringite needles. The lower image, a cement paste after 7 days hydration, illustrates ettringite with its needle-like habit, blocky calcium hydroxide crystals showing their characteristic cleavage parting along the basal plane, a Type I calcium-silicate-hydrate gel (C-S-H) identified by its short needle-like form and and fine bundles of Type I C-S-H, platy-Type II C-S-H, and ettringite needles. Type II C-S-H, more foil-like in appearance. SE imaging is principally applied in the examination of early-age paste microstructure, high-magnification imaging of

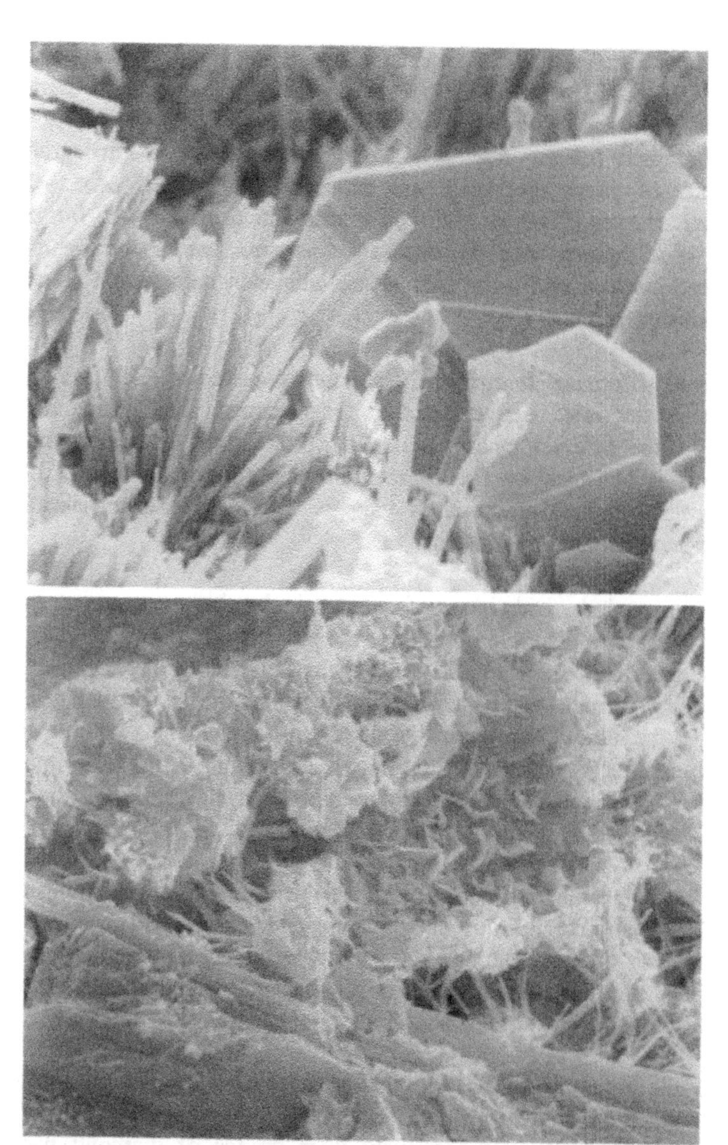

Figure 1. Secondary electron images showing the hexagonal habit of calcium hydroxide, needle-like habit of ettringite, and the sheet-like habit of calcium-silicate-hydrate. In the upper image, crystal growth into a void, where space restrictions are minimal, allowing development of euhedral forms. The lower image, being a more mature paste with limited space, exhibits crystal forms are more subhedral to anhedral. Here one may identify the plate-like CH morphology.

microstructural features, and for examining aggregate texture. Knowledge of the morphological and compositional characteristics of the hardened cement paste constituents is invaluable for their identification. While SE imaging is useful for examining surface texture and crystal morphology, the rough surface makes measurements of phase abundance and distribution unreliable. As the hardened cement paste matures, filling of the void spaces eliminates the well-formed crystals shown in these figures with the resulting microstructure appearing nondescript. Backscattered electron and x-ray imaging are ultimately more useful in examination of these microstructures.

Backscattered Electron Imaging: Backscattered electrons (BE) are high-energy beam electrons that have been scattered by the specimen. The BE image contrast is generated by the different phases' compositions relative to their average atomic number, and is observed by the differential brightness in the image. Anhydrous cement appears brightest followed by calcium hydroxide, calcium silicate hydrate, and aggregate; voids appear dark (Figure 2).

Two morphological forms of calcium hydroxide are common, elongated crystals and massive. The elongated crystals are cross-sections of the hexagonal plates. These exhibit characteristic cleavage parting along basal planes and may represent early-formed CH where relatively unrestricted growth allowed the formation of the ideal hexagonal habit. The term massive CH (without form) may be used where the fine-grained habit shows no crystal form. This material fills voids in the C-S-H, and so, is probably a later-occurring product of the hydration process.

SEM analysis using backscattered electron and X-ray imaging requires a highly polished surface for optimum imaging and X-ray microanalysis. Rough-textured surfaces diminish the image quality by reducing contrast and poor feature definition [2]. Additionally, the lack of a polished specimen makes quantitative estimates arduous, as the surface is no longer planar.

Sample preparation uses an epoxy to permeate the material's pore system. Specimens are then cut or ground to expose a fresh surface, which is then polished using a series of successively finer grades of diamond paste. This polishing stage removes the cutting and grinding damage, exposing a cross section of the material's microstructure. Epoxy impregnation of the pore system serves two purposes: A) it fills voids and, upon curing, supports the microstructure, restraining it against shrinkage cracking, and B) it enhances contrast between the pores, hydration products, and cementitious material.

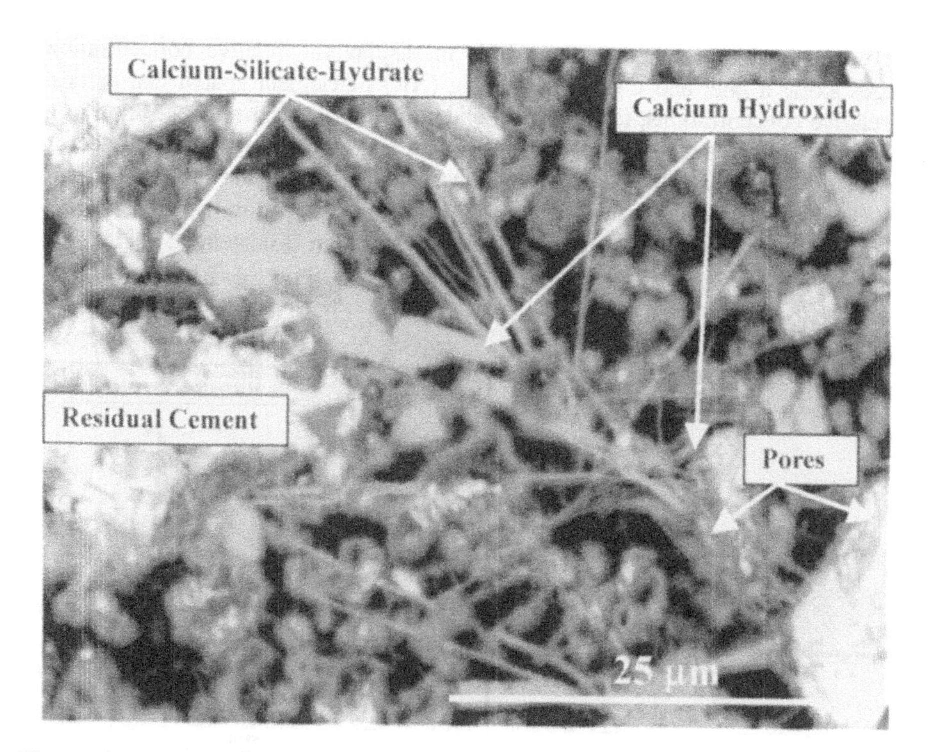

Figure 2. BE imaging of hardened cement paste at 1 day. Identification of calcium hydroxide is made by its relative brightness (between CSH and anhydrous cement), morphological characteristics (cross-sections of the plate-like crystals), and spot chemical analysis (calcium and oxygen). The BE image from a planar cross section, is amenable to both image analysis and x-ray microanalysis.

X-Ray Microanalysis: X-radiation is produced when a specimen is bombarded by high-energy electrons. X-ray microanalysis systems generally employ an energy-dispersive detector with the other detector type being a wavelength detector. The energy-dispersive detector has the advantage of collecting the entire spectrum at one time while the wavelength detector acquires the spectrum sequentially. The X-ray energy level is displayed as the number of counts at each energy interval and appears as a set of peaks on a continuous background (Figure 3). The positions of the peaks are characteristic of a particular element, so identifications are made by examination of peak positions and relative intensities. The X-ray signal can be used for: A) spectrum analysis to determine which elements are present and in what concentration; B) line scan analysis to display the relative concentration changes along a line; and C) X-ray imaging (XR) of element spatial distribution and relative concentrations, to aid in phase identification. Mass concentration to a few tenths of a percent can be detected using an energy dispersive X-ray detector with relative accuracy of quantitative analysis (using certified standards) about ±20% for concentrations around 1%, and ±2% for concentrations greater than 50%. More details on x-ray microanalysis may be found in Goldstein et. al [1].

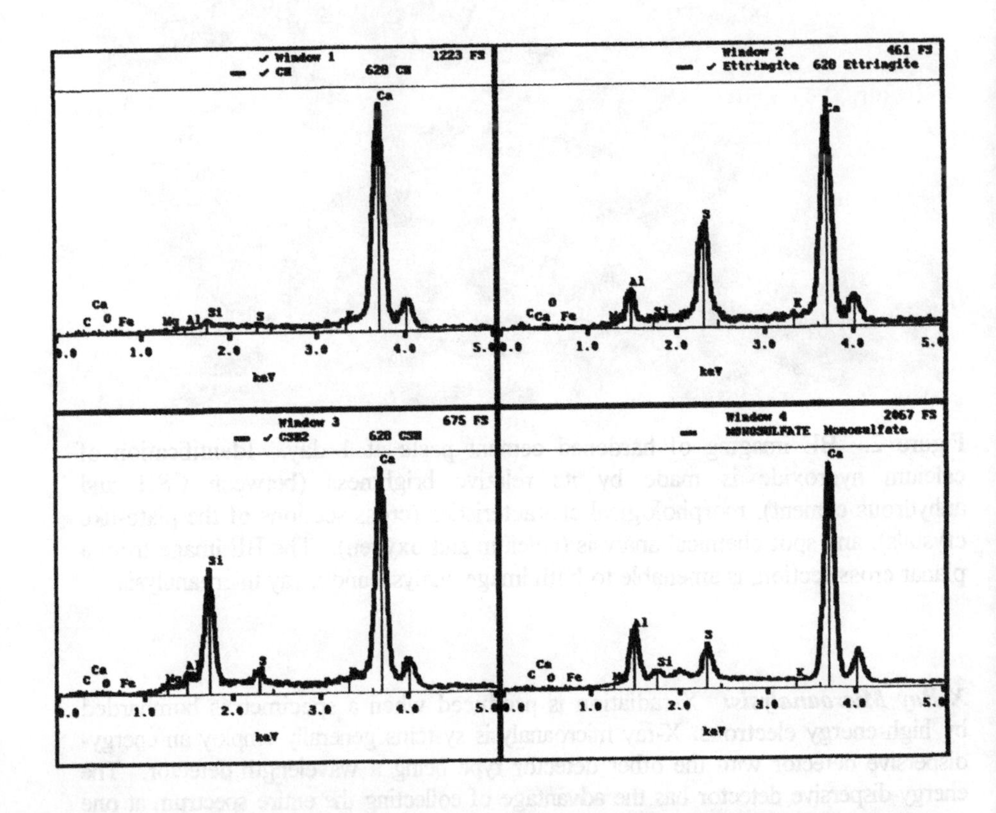

Figure 3. X-ray microanalysis is used for qualitative chemical analysis. These data, with morphological characteristic descriptions, may be used for phase identification through supposition of the element compositions and the relative peak intensities for selected elements.

In Figure 4, BE and X-ray (XR) images display constituents of hardened portland cement paste after about six months hydration. Compared to Figure 2, the large pores are greatly reduced as the hydration products fill the available space. Residual cement grains appear brightest followed the hydration products calcium hydroxide (CH), calcium-silicate-hydrate (CSH), and the darkest being the epoxy-filled pore spaces. Other constituents such as ettringite and monosulfate appear similar in gray level to that of CSH and the use of X-ray analysis and imaging simplifies their identification and imaging their distribution. Both of these phases do show greater uniformity and are a slightly darker gray than that of CSH and often exhibit a platy parting. Feature resolution is dependent upon instrument operating conditions, imaging mode and phase density. Secondary electron imaging resolution may approach nanometer-size while image resolution for the BE is approximately 0.25 μm and, for the x-ray images, about 1 μm.

Preparation of Cement Paste, Mortar, and Concrete Sections

Cement pastes, mortars, and concretes may be prepared in two ways: A) dry potting and B) wet potting. Dry potting is used when the specimen has been dried before, when drying shrinkage-related cracking is not of concern, or when a rapid preparation is needed and utilizes vacuum to permeate the specimen with epoxy. Wet potting is used to prepare a polished section where the material has not been dried and therefore has not undergone any drying shrinkage. Wet potting is a three-step process where the pore solution is replaced with alcohol (200 proof ethanol), then the ethanol is replaced with a low-viscosity epoxy, and then the epoxy is cured. Cracks observed using this preparation may then be ascribed to physical or chemical processes acting upon the concrete, and not due to drying-related shrinkage [3]

A diamond blade slab or a wafering saw, lubricated using propylene glycol exposes a fresh surface that is smoothed by grinding using silicon carbide abrasive papers of 320, 400 and 600 grit. Polishing removes the damage imparted by the sawing and grinding operations using a sequence of successively finer particle size diamond polishing pastes from 9 μm to 0.25 μm, and a lap wheel.

A thin coating of carbon serves to dissipate excess charge from the specimen while exhibiting little effect on image contrast and little interference with elements of interest. Metal coatings such as gold or gold palladium are suitable for secondary imaging of topographic features. However, their x-ray lines interfere with elements of interest and their use decreases BE contrast.

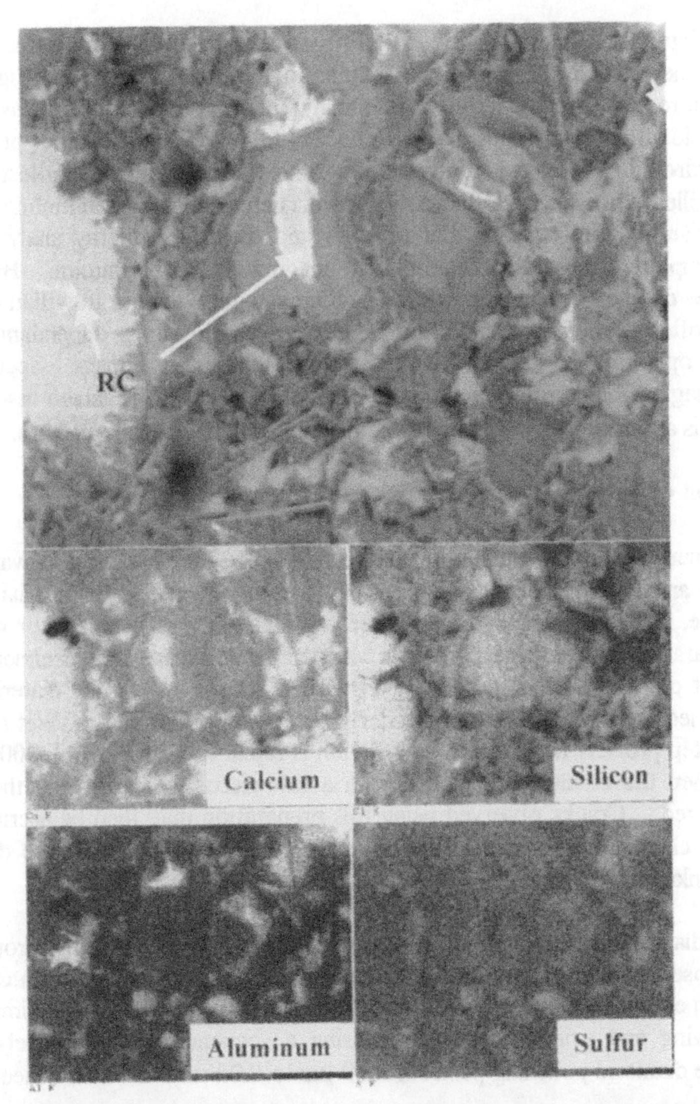

Figure 4. BE and X-ray imaging of a polished section of hardened portland cement paste. Residual cement (RC) appears brightest followed by calcium hydroxide (CH), calcium-silicate-hydrate (CSH), monosulfoaluminate (Afm), and other other hydration products like ettringite and monosulfate. X-ray imaging facilitates distinction of the individual phases through examination of their constituent elements. Field width: 73 micrometers.

Quantifying microstructural features

The relationship between area fraction and volume fraction has been recognized for a long time, and was mathematically derived by Chayes [4] and explained by Galehouse [5] in the following way:

"In a randomly chosen area of a rock such as may be represented by a thin or polished section, the ratio of the area of a particular mineral to the area of all the minerals is a consistent estimate of the volume percent of the mineral in the rock. In simplified mathematical form, let

X = a particular mineral species in thin-section,
H = the thickness of the thin-section,
V = volume,
A = area.

Then the volume of X in the thin-section, V_x, equals $A_x \cdot h$ and the total volume, V, equals $A \cdot h$. Therefore:

$$\frac{V_x}{V} = \frac{A_x \times h}{A \times h} = \frac{A_x}{A} \quad (1)$$

The problem of finding the volume percent of various minerals in a rock can therefore be reduced to finding the area percent of the mineral in a section."

Three approaches in determining the area percent of a constituent in a polished section are; A) the Delesse method of tracing and weighing each phase group; B) the Rosiwal-Shand linear traverse method; and C) the Glagolev-Chayes point count method. BE images are amenable to image analysis, a procedure comparable to that of the Delesse approach, while the Glagolev-Chayes point count method is often simpler and potentially faster. As the name implies, this technique involves sampling the polished section using a grid of points, and identifying the phase falling underneath each point. Additional points are measured by moving the specimen to a new field of view and again counting the phases under the points. The point count method is relatively simple yet provides a reliable means to physically measure phase abundance.

Hofmänner [6] indicated three sources of error that must be considered in point counting:

a) random sample errors,
b) errors in identification by the observer, and
c) measurement errors.

Implicit in this procedure is that a representative sample be collected. The measurement error of the analysis is related to the total number of points counted, with the absolute error at the 96% confidence interval given by:

$$\delta = 2.0235 \times \sqrt{\frac{P \times (100 - P)}{N}} \tag{2}$$

Where

δ = the absolute error in percent,
P = the percentage of points of a phase, and
N = the total of all points.

The total error becomes smaller as the number of points counted increases, decreasing the uncertainty of the measurement. If desired, the results may be expressed as mass percentages, calculated by multiplying the volume fractions by the specific gravity of the corresponding phase and normalizing the totals to 100%

Image Processing and Analysis: Another approach to quantifying phase abundance is through image processing and analysis. Often features of interest may be uniquely identified based upon the gray level of the SEM BE image or by some combination of BE gray level and combination of X-ray images. The gray-level histogram is a plot of the number of pixels in each gray level and is helpful in setting thresholds for the upper- and lower-bounds of gray for a specific phase. The image processing operations, utilizing arithmetical and logical operations on the BE and X-ray images, serve to isolate features of interest while the analysis operations perform a measurement on the isolated features.

In Figure 5, X-ray imaging serves to identify the occurrence and distribution of calcium hydroxide in a hardened cement paste. Processing manipulates the image sets to produce a binary image of CH distribution. Here, CH is identified through specification of regions of moderate BE and moderate calcium intensity. Regions of the images satisfying these criteria are shown in the binary image. This image may then be analyzed to estimate area fraction, phase shape, and distribution.

In Figures 6 and 7, the interfacial transition zone microstructure is assessed using SEM/BE imaging and image analysis. Four constituents are identified using the BE gray levels in Figure 6; residual cement being the brightest followed by calcium hydroxide, C-S-H gel, and pores showing as black. A gray-level histogram to the right of the image plots the number of image pixels across the gray scale. In Figure 7, the image is segmented into the four phases on the basis of gray level, and the constituents area fractions are estimated as a function of distance from the interface. This provides a graphical display of the changes in paste microstructure with distance. This plot is scaled to fit the image and is based upon single 10 μm-wide field estimates from the interface so no uncertainty values are available.

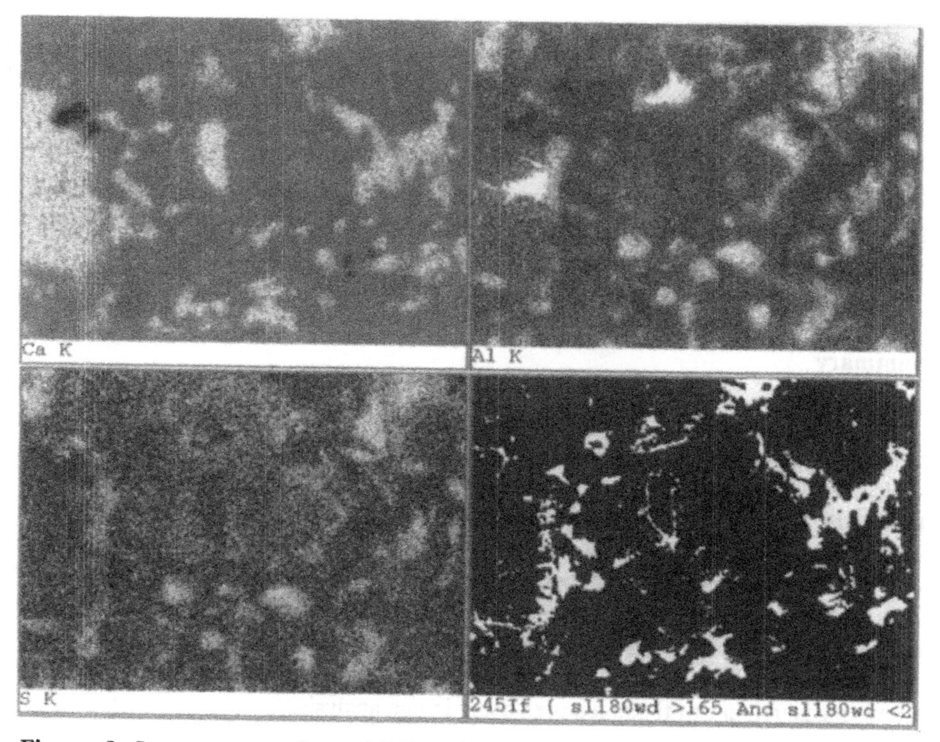

Figure 5. Image processing of BE and calcium X-ray images here highlights calcium hydroxide location. Analysis of the binary image of CH distribution (lower-right image) estimates 12% of the image field to be occupied by calcium hydroxide.

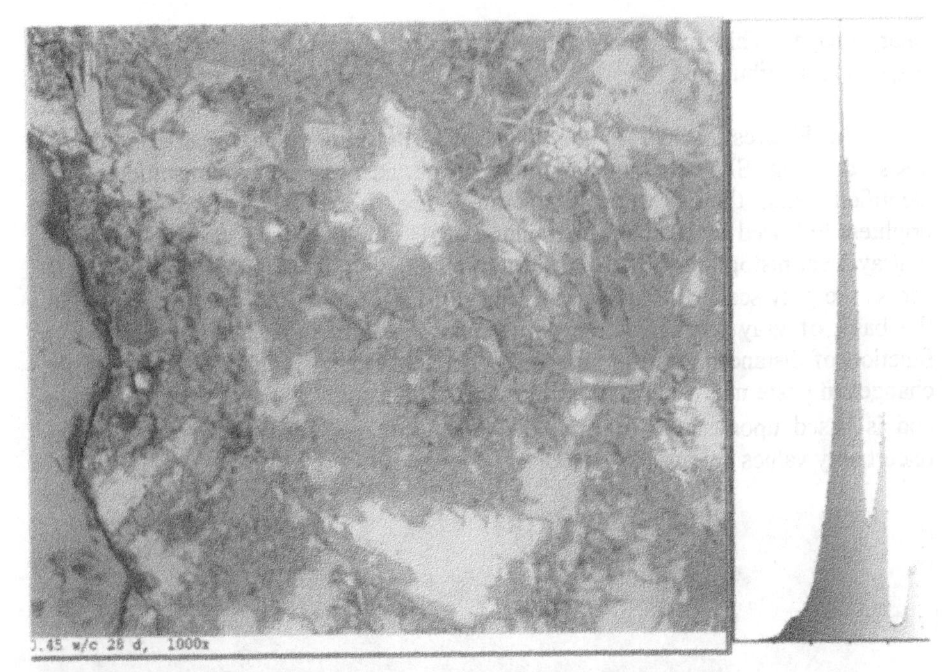

Figure 6. The paste / aggregate interfacial transition zone often shows increased porosity (dark) and calcium hydroxide, and less residual cement than the bulk cement paste. 150 μm field width.

Summary

The application of scanning electron microscopy enhances our ability to characterize cement and concrete microstructure, and will aid in evaluating the influences of supplementary cementing materials, evaluating concrete durability problems, and in the prediction of service life. SEM and X-ray imaging techniques allow imaging the complex microstructure of concrete with sub-micrometer definition. Careful specimen preparation is important as both BE and XR imaging require highly polished surfaces. Through combination of the information available in the backscattered electron and X-ray images, an accurate segmentation of the image into its constituent phases may be achieved. This allows measurement of those features using image analysis.

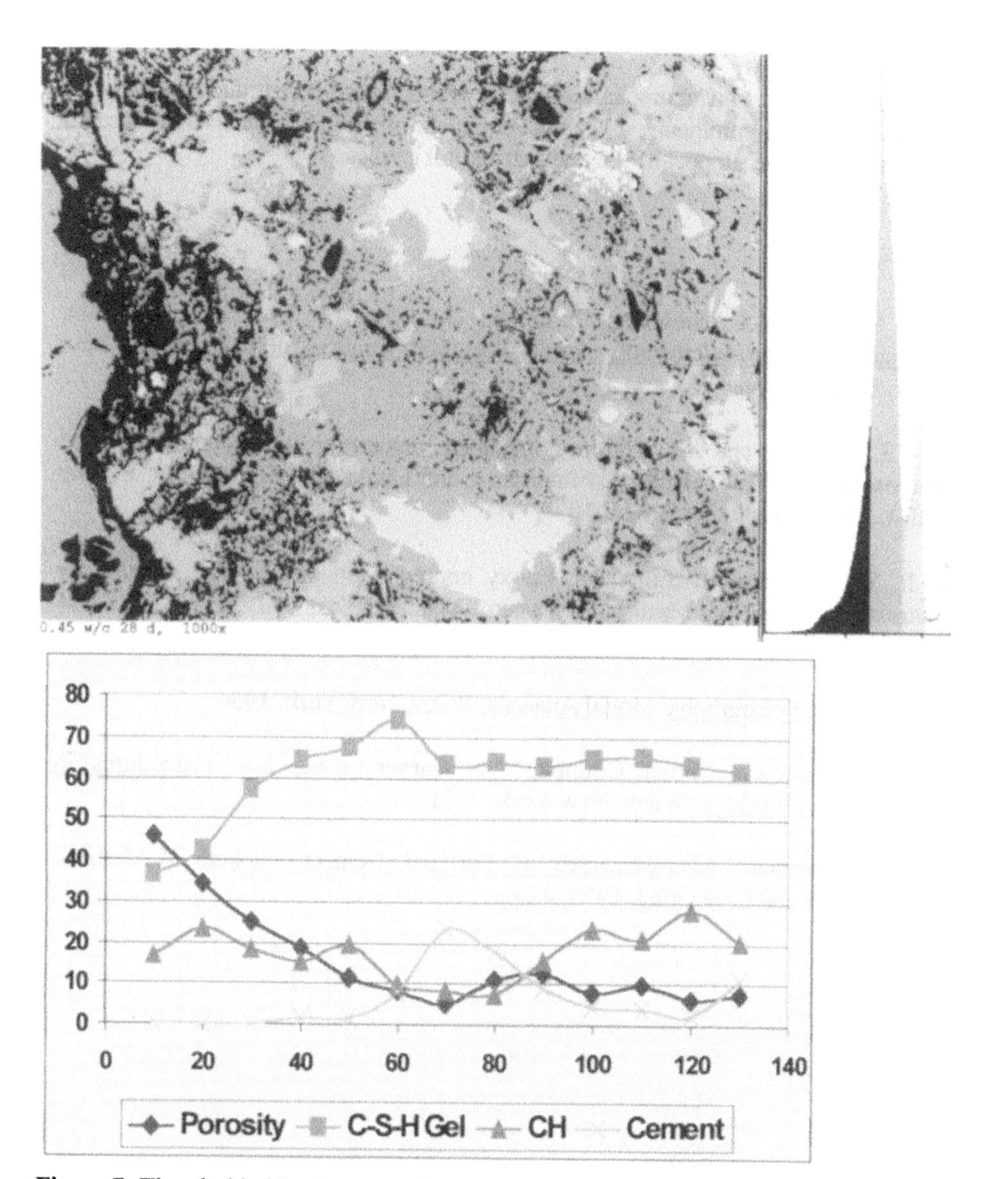

Figure 7. Thresholded backscattered electron image and gray-level histogram of interfacial transition zone allows identification and plot (to scale) of lateral distribution of porosity, C-S-H gel, calcium hydroxide, and residual cement: 150 μm field width.

Acknowledgements
The author extends his appreciation to reviewers Chiara Ferraris and Ed Garboczi of NIST for their comments and suggestions, and the Building and Fire Research Laboratory for the support of the microstructural studies.

References

1. J. Golstein, D.E. Newbury, P. Echlin, D.C. Joy, C. Fiori, and E. Lifshin, Scanning Electron Microscopy and X-Ray Microanalysis, Plenum Press, New York, 1981, 673 pp.

2. P.E. Stutzman and J.R. Clifton, "Specimen Preparation for Scanning Electron Microscopy," in Proc. of the 21st Internat. Conf. On Cement Microscopy, L. Jany and A. Nisperos eds., pp. 10-22, 1999.

3. L. Struble and P.E. Stutzman "Epoxy impregnation of hardened cement for microstructural characterization," Journal of Materials Science Letters, 8, 632-634 (1989).

4. F. Chayes, Petrographic Modal Analysis, Wiley, New York, 1956.

5. J.S. Galehouse, "Point Counting", in Carver, R.E., Ed., Procedures in Sedimentary Petrology, Wiley, New York, 1971.

6. F. Hofmanner, Microstructure of Portland Cement Clinker, Holderbank Management and Consulting, 1973, 48 pp.

AN OVERVIEW OF THE ROLES OF $Ca(OH)_2$ IN CEMENTING SYSTEMS

Paul W. Brown
Penn State University

Boyd A. Clark
RJ Lee Group, Inc.

ABSTRACT

This paper is an overview of the role $Ca(OH)_2$ plays in the CaO-SiO_2-H_2O system. The paper will examine the role of $Ca(OH)_2$ in two broad areas: 1) in the hydration of C_2S and C_3S, and 2) in the durability of this cement system. The hydration reactions can be described by the following equations:

$$3CaO \cdot SiO_2 \xrightarrow{H_2O} 1.7CaO \cdot SiO_2 \cdot 4H_2O + 1.3Ca(OH)_2$$

$$2CaO \cdot SiO_2 \xrightarrow{H_2O} 1.7CaO \cdot SiO_2 \cdot 4H_2O + 0.3Ca(OH)_2$$

These reactions all occur at the C-S-H-$Ca(OH)_2$ invariant point, observed in the CaO-SiO_2-H_2O diagram.

The CaO-SiO_2-H_2O diagram provides a method to determine the phases formed. A review of this phase diagram shows many important aspects; calcium silicate hydrate (C-S-H) is observed in a distinct phase field of variable composition, the calcium to silicon (C/S) ratio of C-S-H varies between 0.83 and 2.0.

Hydrated compounds form through dissolution and re-crystallization. The formation of solids can be graphically represented with respect to the phase diagram, the C_3S or C_2S composition and the compositional point for pure water. If the line of congruent dissolution of either C_3S or C_2S intersects the C-S-H solubility curve then 1) C-S-H will be the first-formed solid and 2) the first-formed C-S-H will have a C/S ratio greater then the value at the solubility line point of intersection. If the lines of congruent dissolution of C_3S and C_2S do not intersect the C-S-H solubility curve than hydrous silica will be the first-formed solid and because the C/S ratio in the C-S-H formed is < 0.83. $Ca(OH)_2$ will be also be a hydration product.

$Ca(OH)_2$ formation does not occur for several hours, even in OPC where the common ion effect on OH^- is significant. In the hydration of C_3S and C_2S, $Ca(OH)_2$ formation initially

occurs under conditions where the C/S ratio of C-S-H may exceed that of the C-S-H that forms at the C-S-H-$Ca(OH)_2$-solution invariant point.

Studies of the nucleation and hydration kinetics of C_3S indicate that hydration can be accelerated or retarded by inoculating C_3S suspensions with various materials. General observations from these studies indicate that relatively inert surfaces have little or no effect on C_3S hydration, while active materials tend to accelerate or retard. An important exception is that the effect of added $Ca(OH)_2$ is minimal.

The role $Ca(OH)_2$ plays in durability is influenced by the microstructural aspects of its formation. $Ca(OH)_2$ crystals observed in hardened concrete are large. This suggests that $Ca(OH)_2$ nucleation is infrequent, but once nucleated, $Ca(OH)_2$ crystallites grow relatively rapidly. In concrete, $Ca(OH)_2$ and C-S-H are not distributed homogeneously throughout the paste. $Ca(OH)_2$ forms preferentially in association with aggregate. Implications of this distribution are: $Ca(OH)_2$ does not play a significant role in influencing hydration kinetics by acting as a physical barrier, $Ca(OH)_2$ is important in affecting properties which rely on the transfer of stress from the paste to the aggregate and the depletion of $Ca(OH)_2$ from the interfacial zone may significantly increase the permeability of concrete.

$Ca(OH)_2$ is the most soluble hydrated solid phase normally present in mature concrete. Reaction of $Ca(OH)_2$ with various species to form solids having solubilities less than that of $Ca(OH)_2$ can lead to the formation of AFm and AFt. Multi-component phase diagrams best describe these stability relationships.

Ettringite formation can be described with reference to the model system: CaO-$CaSO_4$-Al_2O_3-H_2O. In the stable system ettringite co-exists with C_3AH_6. In the metastable system, it co-exists with AFm. The metastable system is more relevant to concrete. In the metastable system, ettringite can co-exist with up to two other solid phases including; ettringite-AFm-$Ca(OH)_2$, ettringite-gypsum-$Ca(OH)_2$, ettringite-gypsum-$Al(OH)_3$ and ettringite-AFm-$Al(OH)_3$. It is processes involving the solids at these invariant points that affect concrete durability. Using the metastable phase diagram and scenarios involving the ingress of sulfated groundwater into concrete systems the formation of ettringite can be evaluated. Various formation paths include formation of ettringite from AFm and sulfate, ettringite from gypsum and $Ca(OH)_2$ provided alumina is available, ettringite from AFm and alumina provided calcium and sulfate are available, and ettringite from hydrous alumina and gypsum provided calcium is available.

The metastable phase diagram can also be used to evaluate whether ettringite formation can occur in the presence of pozzolans. Two circumstances can be explored; when solid $Ca(OH)_2$ is present and when solid $Ca(OH)_2$ has been depleted. These two circumstances are represented on the metastable phase diagram as traversing either the $Ca(OH)_2$ crystallization surface or the hydrous alumina crystallization surface. This analysis indicates that under set conditions ettringite can form whether or not solid $Ca(OH)_2$ is present. At best, only the rate of sulfate attack will be affected if $Ca(OH)_2$ has been depleted, thermodynamics indicates that sulfate attack can occur regardless.

CONCLUSIONS

A review of the CaO-SiO_2-H_2O and CaO-$CaSO_4$-Al_2O_3-H_2O metastable phase diagrams were used to examine the role of $Ca(OH)_2$ in cementing systems. Important aspects determined include C-S-H must have a variable composition, $Ca(OH)_2$ forms separately from C-S-H, microstructural aspects of $Ca(OH)_2$ formation influences durability in concrete, and ettringite formation can occur whether or not $Ca(OH)_2$ is present in the cementing system.

Coffee break: Juri Gebauer

Magdalena and Jan Skalny with Hal Taylor

CHLORIDE BINDING TO CEMENT PHASES: EXCHANGE ISOTHERM, ^{35}Cl NMR AND MOLECULAR DYNAMICS MODELING STUDIES

R. James Kirkpatrick
Department of Geology
University of Illinois at Urbana
Champaign, Urbana, IL 61801, USA

Ping Yu[1]
Department of Materials Science and Engineering
University of Illinois at Urbana
Urbana, IL 61801, USA

Andrey Kalinichev
Department of Geology
University of Illinois at Urbana
Urbana, IL 61801, USA

[1]Current Address: Department of Land Air and Water Resources, 272 Hoagland Hall, University of California at Davis, 1 Shields Avenue, Davis, CA 95616, USA

ABSTRACT

Exchange isotherms, ^{35}Cl NMR spectroscopy, and molecular dynamics modeling probe different time and distance scales relevant to understanding the binding of chloride in cement systems. This paper describes an integrated study of chloride binding to single-phase cement materials using these techniques. Calcium hydroxide shows significant chloride exchange and provides a useful reference point for understanding the chloride exchange behavior of C-S-H and AFm phases.

INTRODUCTION

Binding of chloride to the hydration products in cement paste is thought to substantially decrease the rate of chloride penetration into concrete and to increase the time needed for it to participate in depassivation of the protective oxide layer of the reinforcing steel. Previous workers have distinguished three types of chloride in cement pastes: free (solution), sorbed (bound), and lattice (structural) [1-6]. Sorption is thought to both retard the net rate of transport in concrete and to provide a reservoir of chloride for potential detrimental damage [3,7]. The mechanisms of chloride binding and even the binding capacities of the relevant phases are not well understood. This paper presents a brief summary of a combined exchange (sorption) isotherm, ^{35}Cl NMR and molecular dynamics

(MD) modeling study of chloride binding to individual hydrated cement phases designed to better understand the structural-chemical factors that control the chloride exchange capacities of the phases present in hydrated cement paste. Sorption isotherm data provide a macroscopic picture of the total binding capacity of a given phase, and modeling of the data with multi-layer theory can provide insight into the sorption mechanisms [8]. The ^{35}Cl NMR techniques provide a measure of the total binding capacity and insight into the structural environments of the sorbed chloride and its molecular scale dynamical behavior, significantly improving understanding of the sorption isotherm data [9,10]. MD models provide a more specific molecular-scale picture of the surface sorption sites and the dynamics of the motion of chloride at and near the surface and thus further basis for interpreting the isotherms and NMR data [11,12]. Somewhat surprisingly, calcium hydroxide (portlandite) exhibits significant chloride exchange capacity. It provides a useful starting point to understand the chloride exchange behavior of C-S-H and AFm phases.

EXPERIMENTAL AND COMPUTATIONAL METHODS

The experimental and computational methods used are described in references 8 – 12. Briefly, the exchange isotherm studies were done at room temperature by equilibrating portlandite, the carboaluminate AFm phase ($C_4A\overline{C}H_{11}$), ettringite, jennite, C-S-H samples with analyzed C/S ratios of 1.5 and 0.9, and aluminous C-S-H samples with C/(S+A) ratios of 0.8 and 1.36 with lime saturated sodium chloride solutions. Solid/solution ratios were 0.2g/25ml, the chloride concentrations varied from 0.005 to 0.1 M, and the pH was about 12.4. Exchange isotherms were calculated using the standard technique of monitoring the change in solution concentration before and after exchange. The ^{35}Cl NMR experiments were conducted at H_0 = 11.7 T using suspensions of the individual phases in NaCl or KCl solutions and were monitored for chemical shift and T_1 and T_2 relaxation rates. Solid/solution ratios varied to provide stable suspensions. Comparison with the results for neat solutions and suspensions under variable composition and temperature conditions allow investigation of the behavior of the bound chloride. Experiments using variable NaCl solution concentration were conducted at room temperature for portlandite, $C_4A\overline{C}H_{11}$, and jennite. Variable temperature experiments (temperature between 0 and 60 °C) were conducted in 1 M KCl solutions for all the phases listed above.

Molecular dynamics modeling was undertaken for neat 0.25 M NaCl solution, portlandite, the chloroaluminate AFm phase (Friedel's salt), ettringite and tobermorite (model C-S-H) using systems consisting of several unit cells of the crystal and a layer of 0.25 M NaCl solution 2 to 3 nm thick. This thickness for the solution layer is large enough to effectively exclude direct influence of one interface on another under the periodic boundary conditions used, and the number of water molecules in the layer was chosen to reproduce the density of bulk liquid water under ambient conditions (~1 g/cm^3). The interaction parameters were modified from the augmented ionic consistent valence force field (CVFF) within the Cerius^2 molecular modeling package (Molecular Simulations Inc., 1998) [13]. Water molecules were modeled using a flexible version of the simple point charge (SPC) interaction potential. Power spectra of atomic motions in the translational, librational, and vibrational frequency ranges and diffusion coefficients for bulk and surface bound chlorides were calculated from the Fourier transforms of the velocity autocorrelation functions.

CHLORIDE BINDING TO PORTLANDITE

The results of the exchange isotherm experiments show that portlandite has the largest specific exchange capacity (Cl atoms/nm^2) of all the phases investigated, exceeding that of even $C_4A\bar{C}H_{11}$ at high solution concentrations (Figure 1). The extent of exchange increases with increasing solution concentration, never reaches a plateau, and is thus well fit with the Freundlich-type isotherm shown. All ^{35}Cl NMR signals observed for the suspensions are quite narrow, symmetrical peaks (FWHH < 200 Hz). Their widths decrease with increasing solution concentration. Thus, all the observed chloride is in rapid exchange (ν_{ex} > 2 kHz) between the solution and sorbed states. The chemical shifts are within a ppm of the 1 M NaCl solution standard set at 0 ppm, indicating a predominantly hydrated, solution-like local structural environment. In all cases the observed T_1 and T_2 are identical, as expected for rapid exchange conditions. The observed T_1 relaxation rates ($R_1 = 1/T_1$) decrease with increasing solution concentration at constant solid/solution ratio (0.70 g/g for portlandite) and approach the value of the neat solution at high concentrations (Figure 2). T_1 relaxation rates are well known to increase for species near water-solid interfaces due to decreased rates of molecular reorientation [14], and these observations are readily interpreted to indicate that a progressively decreasing fraction of the chloride in each sample is associated with the surface as solution concentration increases. The data can be interpreted quantitatively from the relationship $R_{obs} = (1-□)R_F + □R_B$, where R_{obs}, R_F and R_B are the relaxation rates of the suspension, neat solution and sorbed species; and □ is the fraction of the total chloride that is sorbed [9,15]. R_{obs}, and R_F are experimentally determined (Figure 2), and R_B can be determined by extrapolation to extreme dilution (here 98.4 Hz for portlandite, about three times the value for the neat solution). For portlandite the calculated □-values decrease with

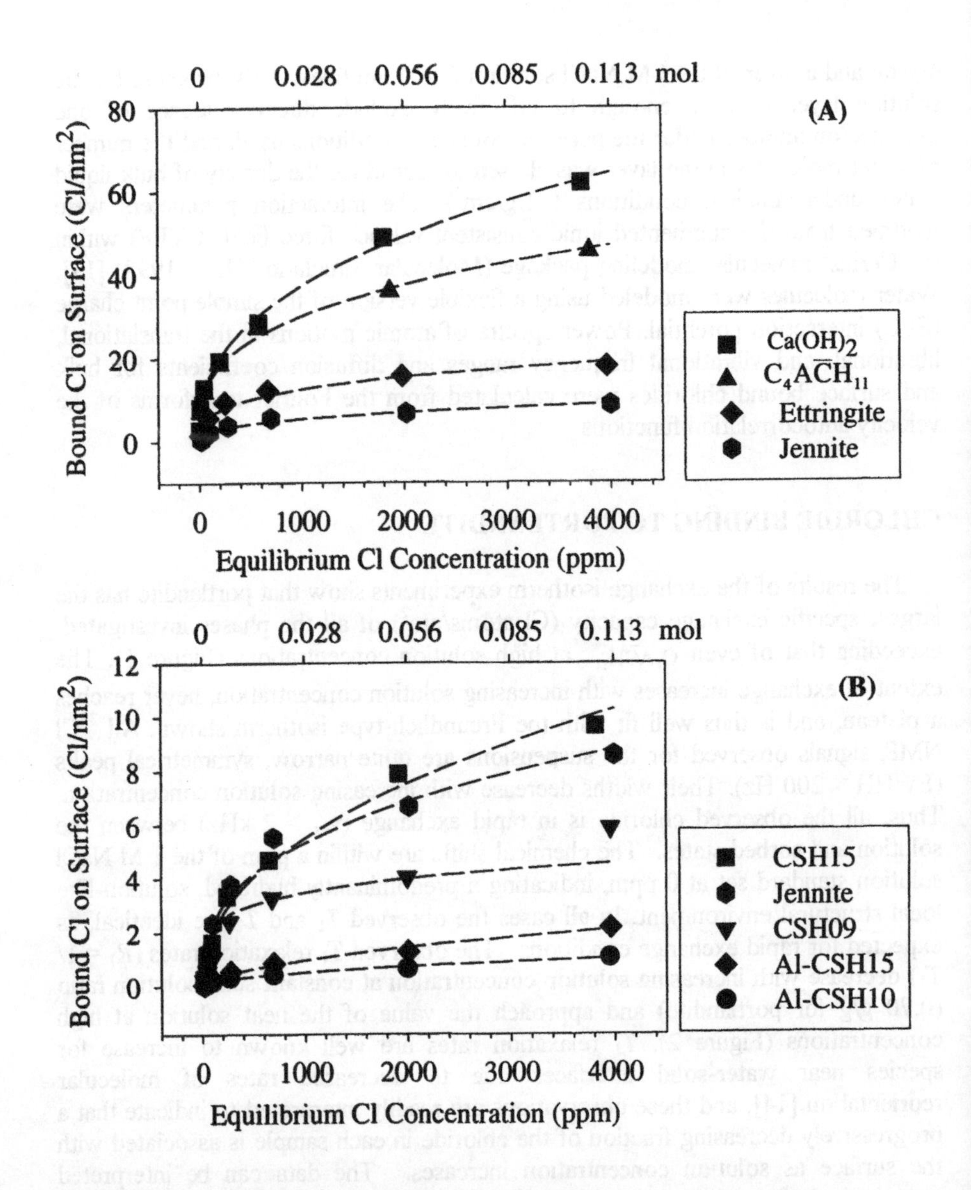

Figure 1. Chloride sorption isotherms for cement hydrate phases in Ca hydroxide saturated 0.25 M NaCl solutions. Fits are to a Freundlich-type isotherm with the relationship $\theta = A \cdot C^{1/n}$, where θ is the sorption density (Cl atoms/nm^2 of solid surface), A is a fitting constant, C is the equilibrium solution concentration, and n is a fitting constant. [After Reference 8]

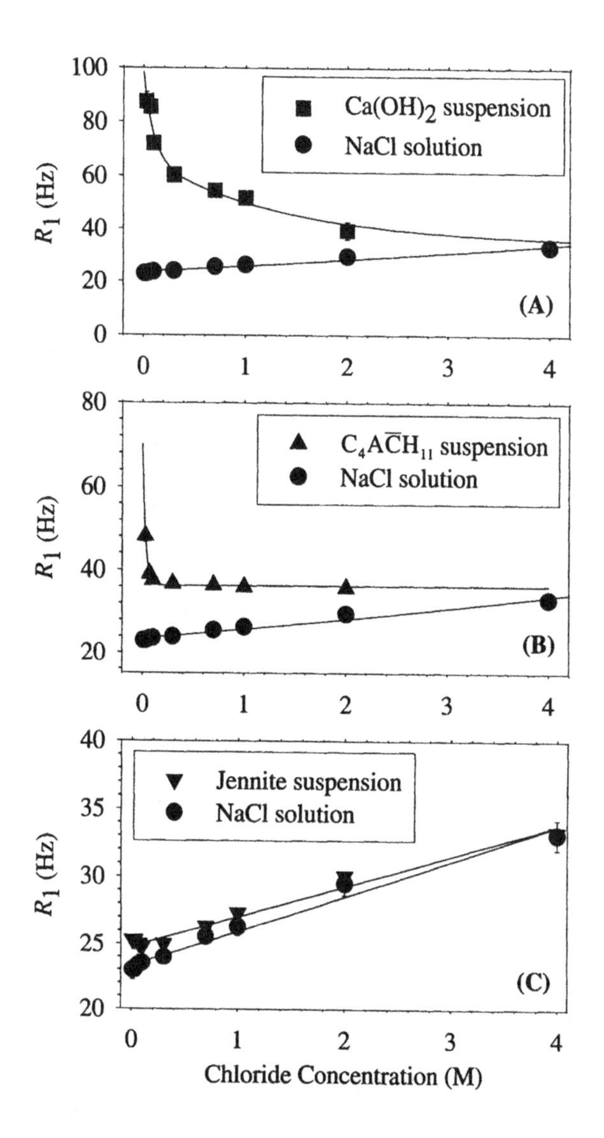

Figure 2. Observed relationships between the ^{35}Cl NMR T_1 relaxation rate ($R_1 = 1/T_1$) and solution concentration for suspensions of portlandite, $C_4A\overline{C}H_{11}$, and jennite in Ca hydroxide saturated NaCl solutions, along with data for the neat Ca- hydroxide saturated NaCl solutions. The differences between the two data sets is related to the extent of chloride sorption and is modeled by the relationship described in the text. [After Reference 9]

increasing chloride concentration, in agreement with qualitative expectation. In contrast, the chloride surface density (Cl/nm^2) increases (Figure 3). and the values are in excellent quantitative agreement with the sorption isotherms (Figure 1). The variable temperature ^{35}Cl NMR R_1 results normalized to unit surface area are also in excellent qualitative agreement with these observations, showing that only Friedel's salt (the Cl-containing AFm phase) has a greater effect (Figure 4).

Together, the binding isotherm and ^{35}Cl NMR data provide significant new insight into chloride binding on portlandite. Surface densities are up to 60 Cl/nm^2, equivalent to many monolayers, which would contain between 3 and 9 Cl/nm^2 depending on the atomic packing. The chloride is, however, not tightly bound, but rather is in a solution-like chemical environment and is in rapid exchange with the bulk solution at frequencies > 2 kHz. Such high sorption densities and the solution-like environment suggest that much of the chloride is in the diffuse layer. Electrical double layer calculations assuming a ζ-potential of +20 mV, the largest ever found for portlandite [17,18] and a solution concentration of about 0.1 M yield sorption densities of about 0.1 Cl/nm^2. The large difference between this and the observed values suggests that both sorption closer to the solid surface than the shear plane relevant to the ζ-potential and ion cluster formation (e.g., Ca-Cl clusters) may contribute significantly to the total sorption [19-25]. Ion cluster formation, which can result in significant co-adsorption and increased sorption for a given surface charge is expected to be significant at the high concentrations in the diffuse layer [19-23]. An important conclusion from this work is the need for increased understanding of the structure of the concentrated, high pH solutions relevant to cement chemistry.

Molecular dynamics modeling of chloride binding cannot currently investigate systems larger than a few nm in linear extent, but recent results show that they are effective in studying molecular scale binding behavior at solid-solution interfaces [11,12]. For portlandite [11] they show that the surface can sorb both chloride and counter ions (Na^+ or Cs^+ in our simulations) due to the flexibility of the surface Ca-OH groups (Figure 5). We distinguish three broadly defined types of aqueous species (Figure 6): 1) inner-sphere surface complexes (coordinated directly to atoms on the solid surface), 2) outer-sphere surface complexes (separated from the surface by one molecular layer of water), and 3) ions in the solution (which in our models feel the presence of the surface but are separated from it by more than one molecular layer of water). In the case of surface-bound Cl^-, surface hydroxyl groups are oriented towards the anion, thus forming a nearly solution-like local environment for the surface-adsorbed Cl^-. On the other hand, in the case of surface-bound Na^+ and Cs^+, the surface OH groups tend to point

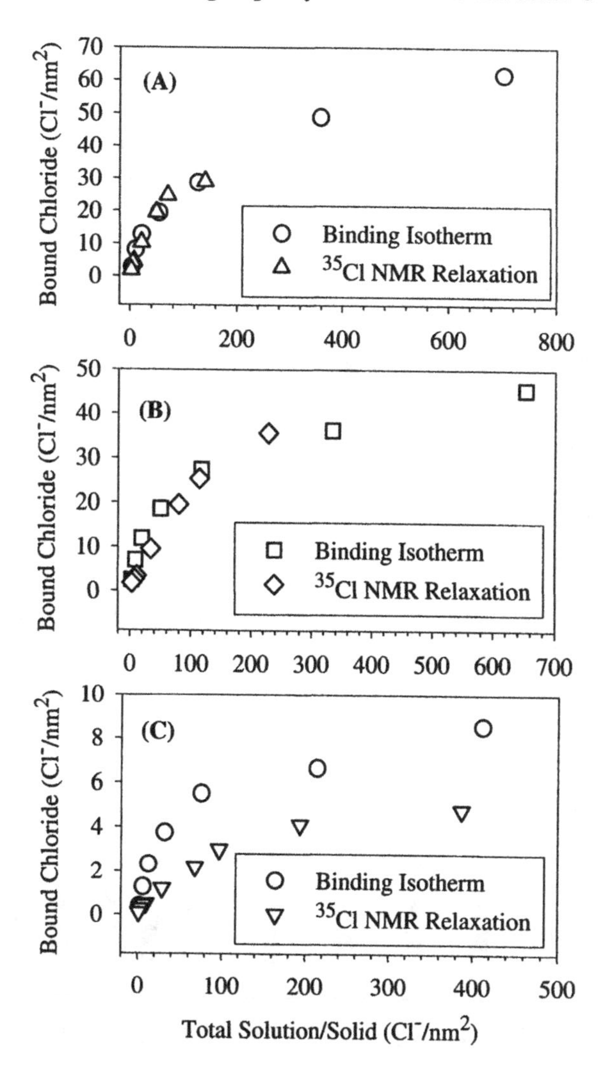

Figure 3. Fraction of sorbed chloride and chloride surface densities for suspensions of A) portlandite, B) $C_4A\bar{C}H_{11}$, and C) jennite determined from the ^{35}Cl NMR relaxation data presented in Figure 2. For portlandite and $C_4A\bar{C}H_{11}$ the sorption densities are in excellent quantitative agreement with the sorption isotherm results in Figure 1. The results for jennite from NMR are about 40% lower, but the densities are so low, that the results are probably within analytical error. [After Reference 9]

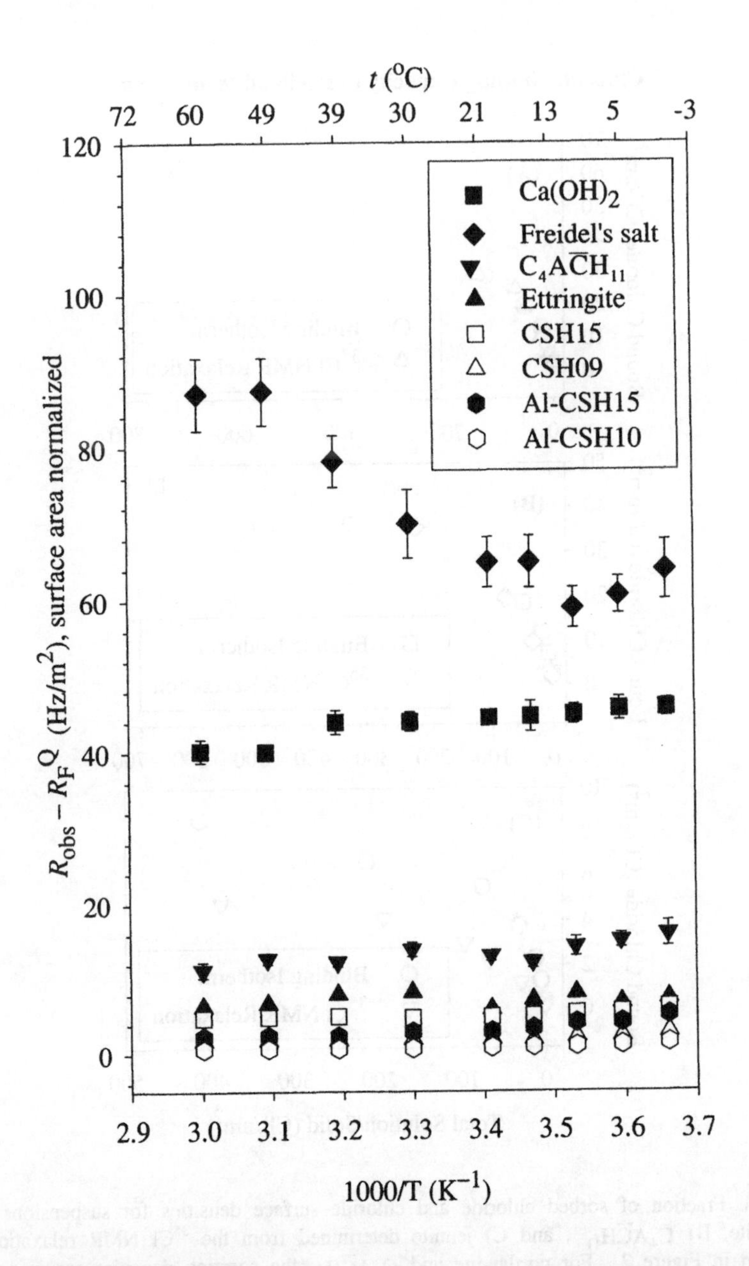

Figure 4. Variable temperature ^{35}Cl NMR T_1 relaxation rates for suspensions of the indicated cement hydrate phases. The relaxation rates presented are $(R_{1,obs} - R_{1,solution})$/surface area, where $R_{1,obs}$ is $1/T_1$ for the suspension, $R_{1,solution}$ is $1/T_1$ for the neat Ca hydroxide saturated 1 M KCl solutions. [After Reference 16]

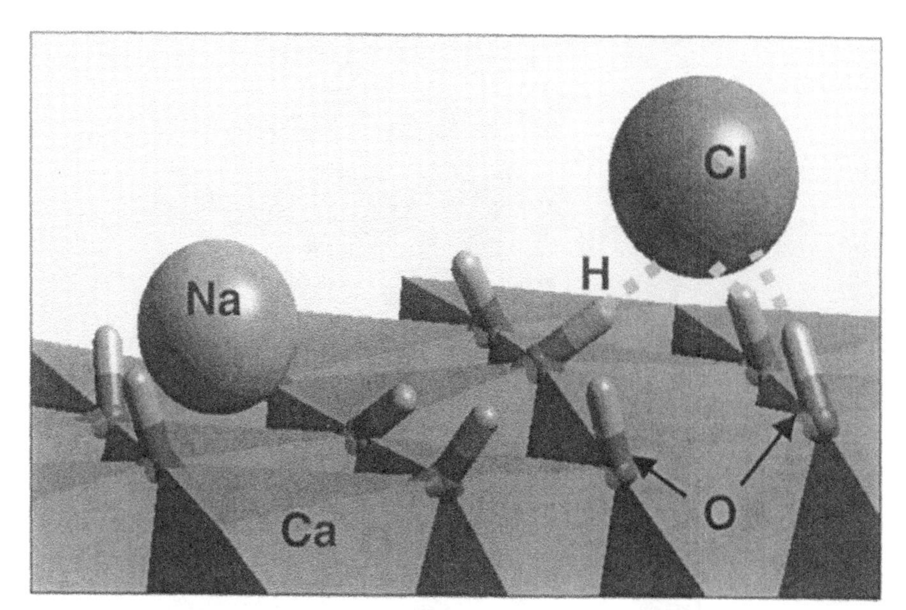

Figure 5. Computer generated picture of the sorption of chloride and sodium ion on the surface of portlandite. For visual clarity, the Ca-octahedra are blue, oxygens are the small red lines, hydrogens are gray, Na is pink, Cl is green. Hydrogen bonds between surface OH-groups and Cl^- are shown as dashed gray lines, and the water molecules are suppressed. Both the Na^+ and Cl^- coordinate directly to the surface sites with no intervening water molecules, forming so-called inner sphere complexes. This direct surface coordination of both cations and anions is facilitated by of the flexibility of the surfaced OH groups. Note that the H-atoms of the OH groups point towards the Cl^- and away from the Na^+. [After Reference 11]

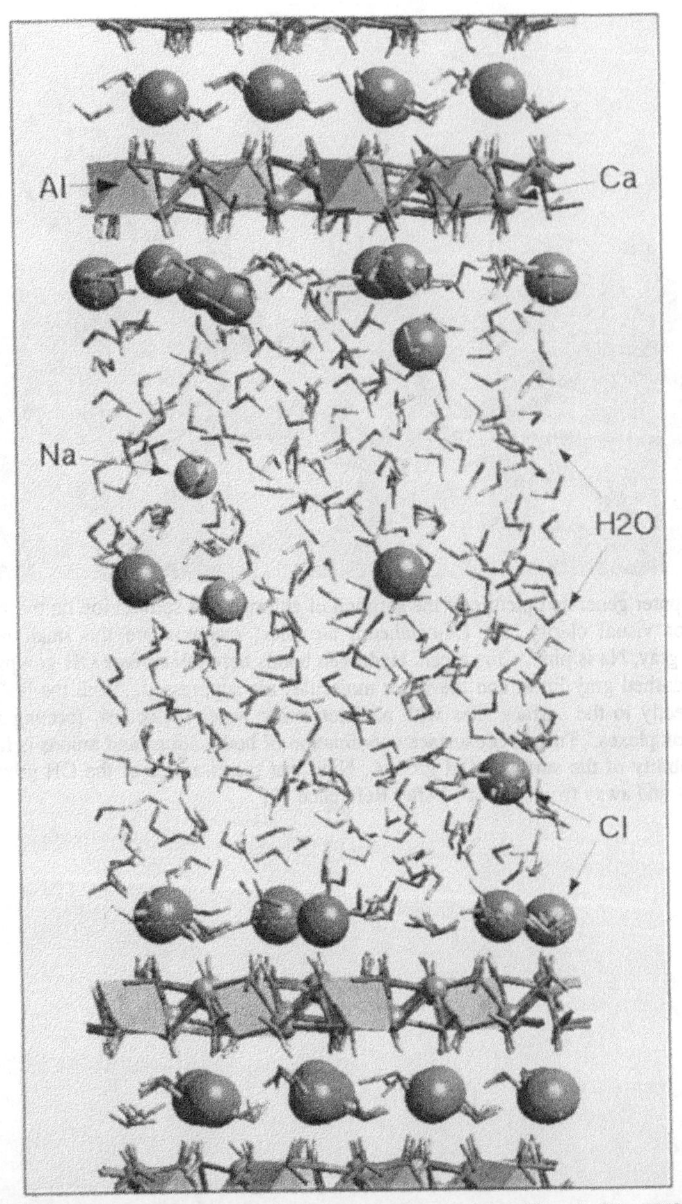

Figure 6. MD-simulation snapshot of the interface between Friedel's salt $[Ca_2Al(OH)_6]Cl\ 2H_2O$ and a ~30 Å-thick layer of 0.25 m aqueous NaCl solution. The crystal structure is represented by the shaded octahedral sheets. Water molecules are the arrowhead shaped symbols. Chlorides and water molecules are ordered in the interlayer, and most surface chlorides occur as inner sphere complexes. [After Reference 11]

away from the cations, thus also creating a favorable electrostatic environments for adsorption by exposing their negatively charged oxygen atoms.

Power spectra of atomic motions allow comprehensive and detailed analysis of the ionic dynamics of the surface species. The spectral density of the low frequency vibrational motions of Cl^- ions consists of two distinct frequency bands centered at approximately 50 and 150 cm^{-1}. Because each chloride ion is H-bonded to neighboring H_2O molecules and/or OH-groups, these two O···Cl···O modes are analogous to, respectively, the intermolecular O···O···O bending and stretching motions of water molecules in the H-bonded network of bulk water [11,12]. The shape of the low frequency vibrational spectrum of surface-bound chloride, thus, closely resembles that of chloride in a bulk aqueous solution. This is in good agreement with the ^{35}Cl NMR results. The situation is quite similar in the case of the dynamics of surface-adsorbed heavy Cs^+ ions, whereas the power spectra of inner-sphere, outer sphere, and bulk solution Na^+ are all different in the simulations [11].

CHLORIDE BINDING TO C-S-H

The extent of chloride sorption on C-S-H is much less than for portlandite, with a maximum of about 10 Cl/nm^2 (Figure 1). It is lowest for the aluminous samples (maximum of about 2 Cl/nm^2), is larger for the sample with C/S = 0.9 and is largest for the sample with C/S = 1.5 and for crystalline jennite. These observations are consistent with previous reports (1,2,26-28). ^{35}Cl NMR results for jennite give sorption densities about 40% lower than those from the isotherms, but the values are probably within experimental error (Figures 2 and 3). As for portlandite, all the observed chloride is in rapid exchange with the bulk solution at greater than kHz frequencies and the chemical shifts are indicative of a solution–like structural environment. The variable temperature ^{35}Cl NMR R_1 data confirm that sorption on C-S-H phases has a relatively small effect on the relaxation (Figure 4).

These observations can be readily interpreted by reference to the chloride exchange results for portlandite and the structure of synthetic, precipitated C-S-H, which is based on that of tobermorite [24-31]. At low C/S ratios near 0.8, the C-S-H structure is comparable to that of tobermorite, with composite layers consisting of a central sheet of Ca-polyhedral sheet sandwiched between dreirketten chains of silicate tetrahedra. Increasing C/S ratios are accommodated first by omission of some of the so-called bridging tetrahedra in the chains and then, at C/S ratios > 1.3, by omission of chain segments and replacement of them by OH groups on the Ca-sheet. C-S-H is most commonly observed to have negative ζ-potentials [17,18], as expected at the high pH of cement solutions, and suggesting weak chloride sorption. MD modeling of chloride interaction with tobermorite shows that chloride is not attracted to the surface [11], in agreement with this idea. Al is thought to substitute for Si on primarily the bridging tetrahedra [32,33], increasing the negative surface charge due to this +3 for +4 charge substitution. Thus, low chloride exchange capacities are expected at low C/S ratios, and especially for aluminous samples, as observed. The jennite structure is thought to be broadly similar to that of tobermorite, but with every other chain missing and replaced by OH groups, along with significant rearrangement of the Ca-sheet [34]. Thus, it probably exposes surface sites comparable to those of portlandite, consistent with its increased chloride exchange capacity. Based on the structural model for synthetic C-S-H described above, the C-S-H sample with C/S = 1.5 is likewise expected to expose significant numbers of Ca-OH sites, accounting for its comparatively large chloride exchange capacity.

Although the chloride exchange capacities for all the C-S-H samples are lower than for portlandite, they are still significant, and for the jennite and the C/S = 1.5 sample probably more than a statistical monolayer. Here also, co-adsorption due to ion cluster formation may contribute to the total exchange capacity. In addition, ^{17}O MAS NMR results for synthetic C-S-H samples suggest the presence of Ca-OH sites in the structure even at low C/S ratios [30], possibly on the broken layer edges. Such sites may also contribute to the observed chloride exchange capacity.

CHLORIDE BINDING TO AFm AND AFt PHASES

The AFm phases have layer structures with positive charges compensated by interlayer anions and correspondingly positive ζ-potentials [17,18,35]. They are expected to have significant chloride exchange capacities, as observed for $C_4A\bar{C}H_{11}$ (Figure 1). The ^{35}Cl NMR results for $C_4A\bar{C}H_{11}$ yield exchange capacities essentially identical to the exchange isotherm studies (Figure 3), and as for portlandite and the C-S-H phases all the signals are quite narrow, symmetrical peaks near 0 ppm. Thus, the Cl^- is in solution-like environments and is in rapid exchange with the bulk solution. The variable temperature ^{35}Cl NMR results for $C_4A\bar{C}H_{11}$ show a much smaller effect than for portlandite, but a larger effect than for the C-S-H phases (Figure 4). The effect for Friedel's salt, the chloride-containing AFm phase, is the largest for all the phases investigated (Figure 4). The temperature dependence is different (negative slope) than for the other phases, suggesting that its relaxation rates are controlled by large fluctuations in the electric field gradient due to rapid exchange of chloride between inner sphere complexes and solution-like sites [16]. MD models for chloride binding on Friedel's salt [12] show formation of such inner sphere sites and exchange of chloride between them and the solution at frequencies of about 2×10^{-10} Hz, orders of magnitude faster than required to cause the observed single, narrow peak.

Structurally, the AFm phases are related to portlandite by ordered substitution of Al for Ca on 1/3 of the hexagonal sites, resulting in the permanent positive layer charge. The large effect of Friedel's salt on the T_1 relaxation rates is, thus, expected due to its strong affinity for chloride and the MD results suggesting exchange between the surface and solution sites [10,12]. The comparatively low chloride exchange capacity of $C_4A\overline{C}H_{11}$, which is equal to or less than that of portlandite, depending on the solution concentration, is probably due to strong attachment of carbonate on the surface sites, which would reduce the positive surface charge and prevent binding. Carbonate has a charge of –2, compared to –1 for chloride, and is thus expected to bind much more strongly to the surface.

Ettringite (AFt) has a columnar structure with a permanent positive charge that is compensated by intercolumnar sulfate and has positive ζ-potentials [17,18,36]. Its chloride sorption density, however, is less than for $C_4A\overline{C}H_{11}$, and the variable temperature ^{35}Cl NMR T_1 results confirm that it has a significantly lower affinity for chloride than portlandite and the AFm phases. MD models show that if the sulfate ions are in their normal structural positions on the ettringite surface, chloride is not attracted to it and does not exchange with the sulfate. As for carbonate, sulfate has a charge of –2, and is expected to be more strongly attracted to the surface than chloride. In addition, the chloride AFt phase is stable only at low temperatures, <0°C, [35] suggesting that the structural conformation of ettringite is more favorable for sulfate than for chloride.

ACKNOWLEDGEMENTS

This research was supported by National Science Foundation and industrial funding to the Center for Advanced Cement-Based Materials and by NSF Grant EAR 97-05746.

REFERENCES

Ramachandran, V. S. (1971) "Possible states of chloride in the hydration of tricalcium silicate in the presence of calcium chloride", *Materiaux et Constructions,* **4**, 3-12.

Beaudoin, J. J., Ramachandran, V. S, Feldman, R. F. (1990) "Interaction of chloride and C-S-H", *Cem. Con. Res.*, **20**, 875-883.

Diamond, S. (1986) "Chloride concentrations in concrete pore solutions resulting from calcium and sodium chloride admixtures", *Cem. Concr. Aggreg.*, **8** (2), 97-102.

Arya, C., Newman, J. B. (1990) "An assessment of four methods of determining the free chloride content of concrete", *Mater. Structure*, **23** (137), 319-330.

Haque, M. N., Kayyali, O. A. (1995) "Aspects of chloride ion determination in concrete", *ACI Mat. J.*, **92** (9/10), 532-541.

Byfors, K., Hansson, C. M., Tritthart, J. (1986) "Pore solution expression as a method to determine influence of mineral additives on chloride binding", *Cem. Con. Res.*, **16**, 760-770.

Hussain, S. E., Al-Gahtani, A. S., Rasheeduzzafar, (1996) "Chloride threshold for corrosion of reinforcement in concrete", *ACI Mater. J.*, **94** (6), 534-538.

Yu, P., and R. J. Kirkpatrick (2000), "Chloride binding to cement hydrate phases: binding isotherm studies," *Cem. Con. Res.*, submitted.

Yu, P., and R. J. Kirkpatrick (2000), "^{35}Cl NMR relaxation study of Cement Hydrate Suspensions", *Cem. Con. Res.*, submitted.

Kirkpatrick, R. J., Yu, P., Hou, X., Kim, Y. (1999) "Interlayer structure, anion dynamics, and phase transitions in mixed-metal layered hydroxides: Variable temperature ^{35}Cl NMR spectroscopy of hydrotalcite and Ca-aluminate hydrate (hydrocalumite)", *Am. Miner.*, **84**, 1186-1190.

Kalinichev, A. G., and R. J. Kirkpatrick (2000), "Molecular dynamics modeling of chloride binding to the surfaces of portlandite, hydrated Ca-silicate and Ca-aluminate phases", *Cem. Con. Res.*, submitted.

Kalinichev, A. G., R. J. Kirkpatrick, and Cygan, R. T. (2000), "Molecular modeling of the structure and dynamics of the interlayer and surface species of mixed-metal layered hydroxides: chloride and water in hydrocalumite (Friedel's salt)", *Am. Mineral.* , **85**, 1046-1052.

Molecular Simulations Inc. (1999). Cerius2-4.0 User Guide. Forcefield-Based Simulations, MSI, San Diego.

Lindman, B., Forsen, S. (1976) *Chloride, Bromine and Iodine NMR, Physico-Chemical and Biological Applications*, (NMR Basic Principles and Progress), Vol **12**, Diehl, P., Fluck, E., and Kosfeld, R., eds. Springer-Verlag, Berlin, Heidelberg, New York.

Kim, Y., and Kirkpatrick (1998), "^{133}Cs and ^{23}Na NMR T_1 investigation of cation sorption on illite", *Am. Mineral.*, **83**, 661 - 665.

Yu, P., (2000) "*Spectroscopic investigation of cement hydrate phases and their chloride binding properties*", Ph.D. thesis, Department of Materials Science and Engineering, University of Illinois at Urbana-Champaign, 137 p.

Babushikin, V. I., Mokristskaya, L. P., Novikova, S. P., Zinov, V. G. (1974) "Study of physical-chemical processes during hydration and hardending of expansive cements," 6th International Congress on the Chemistry of Cement, Supplementary Paper, Section III-5, Moscow, Sept. 1974.

Neubauer, C. M., Yang, M., Jennings, H. M. (1998) "Interparticle potential and sedimentation behavior of cement suspensions: effects of admixtures", *Advn. Cem. Bas. Mater.*, **8**, 17-27.

Ohtaki, H., Radnai, T. (1993) "Structure and dynamics of hydrated ions", *Chem. Rev.*, **93** (3), 1157-1204.

Enderby, J. E., Cummings, S., Herdman, G. J., Neilson, G. W., Salmon, P. S., Skipper, N. (1987) "Diffraction and the study of aqua ions", *J. Phys. Chem.*, **91**, 5851-5858.

Bounds, D. G. (1985) "A molecular dynamics study of the structure of water around the ions Li^+, Na^+, K^+, Ca^{++}, Ni^{++} and Cl^-", *Molecular Physics*, **54** (6), 1335-1355.

Impey, R. W., Madden, P. A., McDonald, I. R. (1983) "Hydration and mobility of ions in solution", *J. Phys. Chem.*, **87**, 5071-5083

Hunt, J. P., Friedman, H. L. (1983) "Aquo complexes of metal ions", in *Progress in Inorganic Chemistry*, Ed. Lippand, S. J., Volume **30**, An Interscience Publication, John Wiley & Sons, pp 359-364.

Hunter, R. J. (1993) "Introduction to modern colloid science", Oxford University Press, New York.

Lyklema, J. (1983) "Adsorption of Small ions"; pp. 223-319 in *Adsorption from solution at the solid/liquid interface*. Edited by G. D. Parfitt and C.H. Rochester. Academic Press

Delagrave, A., Marchand, J., Ollivier, J.-P., Julien, S., Hazrati, K. (1997) "Chloride binding capacity of various hydrated cement paste systems", *Advn. Cem. Bas. Mater.*, **6,** 28-35.

Francy, O., Francois, R. (1998) "Measuring chloride diffusion coefficients from non-steady state diffusion tests", *Cem. Con. Res.*, **28** (7), 947-953.

Page, C. L.; Vennesland, Ø (1983) "Pore solution composition and chloride binding capacity of silica-fume cement pastes", *Materiaux et Constructions*, **16**,19-25.

Yu, P., Kirkpatrick, R. J., Poe, B., McMillan, P. F. and Cong, X. D. (1999) "Structure of calcium silicate hydrate (C-S-H): near-, mid-, and far-infrared spectroscopy", *J. Am. Ceram. Soc.*, **82** (3), 742-748.

Cong, X. D., Kirkpatrick, R. J. (1996) "^{17}O MAS NMR investigation of the structure of calcium silicate hydrate gel," *J. Am. Ceram. Soc.*, **79**, 1585-92.

Cong, X.-D., and Kirkpatrick, R.J., "^{29}Si MAS NMR study of the structure of calcium silicate hydrate", *Advn. Cem. Based Mat.*, **3**, 133-144 (1996)

Richardson, I. G. (1999), "The nature of C-S-H in hardened cement", *Cem. Con. Res.*, **29**, 1131 – 1147
Faucon, P., T. Carpentier, A, Nonat, and J. C. Petit 91998), "Triple quantum filtered two dimensional 27Al magic angle spinning nuclear magnetic resonance study of the aluminum incorporation in calcium silicate hydrate", *J. Am. Chem. Soc.*, **120**, 12075-12082.
Gard, J. A., Taylor, H. F. W., Cliff, G., Lorimer, G. W. (1977) " A reexamination of jennite", *Am. Mineral.*, **62**, 365-368.
Taylor, H. F. W. (1997) "Cement Chemistry", 2nd Edition, Thomas Telford, London.
Pöllmann, H., Kuzel, H.-J., Wenda, R. (1989) "Compounds with ettringite structure", *Neues Jahrbuch.Miner.Abh.*, **160** (2), 133-158.

ROLE OF ALKALINE RESERVE IN THE ACIDIC RESISTANCE OF CEMENT PASTES

A. Hidalgo, C. Andrade, C. Alonso
Instituto Eduardo Torroja (CSIC), C/Serrano Galvache s/n, 28033 Madrid, Spain

ABSTRACT

In realistic conditions, relevant to high-level radioactive waste repositories, concrete durability is limited by its interaction with clays and granitic groundwater. Groundwater is a very weakly mineralised solution containing diluted acids. In contact with the concrete it will leach the ions in the pore solution and will react with the solid phases that contain calcium (CH, CSH and ettringite) producing a progressive neutralisation of the alkaline nature cement paste. The resistance of hardened cement pastes to chemical attack and physical degradation is a combination of chemical composition (in particular the so-called alkaline reserve of the cement pastes) and microstructural factors. At present, a wide variety of experimental procedures are used to characterise the leaching properties of materials used in waste management (Van der Sloot, 1990). In most of the cases, kinetic studies in material test-tubes have been made, in order to know the chemical and microstructural degradation in function of the time (Van der Sloot, 1990; Adenot et al. 1992; Faucon et al. 1998; Unsworth et al. 1997). In this work two types of test methods have been used, an acid neutralisation test method and a permeability test method. The dissolution-precipitation phenomena of cement pastes have been studied by acid attack in powders with a controlled particle sizes; this accelerated test does not fully reproduce reality, but it enables the different variables to be compared. In order to evaluate cements for use as a component in sealing materials in a disposal for nuclear wastes, the water permeability characteristics of cement mortars have been studied. Thus different types of cement and blending agents are compared, and the role of the alkaline reserve has been analysed. The selection of the most resistant cement paste will be the main step in the definition of the most suitable concrete type to be used in high level radioactive waste repositories.

INTRODUCTION

Degradation due to the leaching of calcium occurs when the hydrates in cementitious materials dissolve into the surrounding water. This may cause a loss of strength. The degradation rate is very slow; however the evaluation of this degradation is very important for such structures as radioactive-waste repositories, where extremely long-term stability (over 10,000 years) is necessary. Due to the long timescales involved in laboratory experiments using groundwaters, accelerated tests have to be used to qualify the different concretes for deep repositories; also, to investigate their behaviour under different conditions. These data are necessary for long term predictions of their performance.

Hydrated cement paste is a porous medium whose solid phases are stable at very alkaline pH levels. The leaching processes of materials that contain cement pastes are a combination of chemical reactions and diffusional transport, and have to be studied thermodynamically and kinetically. CSH gel, together with $Ca(OH)_2$ and alkalies, dominate the observed chemical properties of the aqueous phase in Portland cement and account for its high pH and calcium solubility. Addition of fly ash, slag or other pozzolan will gradually change this phase balance: both reduce the $Ca(OH)_2$ content as they react. On the other hand, one of the requirements for a physically-impermeable matrix is a low porosity; cement degradation will depend on physical factors including the effect of porosity, compressive strength and density as well as on leachant characteristics including the effect of pH, flow rate, temperature and water chemical composition.

In realistic conditions of high level radioactive waste repositories, concrete durability is based on its interaction with clays and granitic groundwaters. In general groundwater is a very weakly mineralised solution containing diluted acids; however its exact composition is site-specific. The leaching processes of materials that contain cement pastes are a combination of chemical reactions produced by a progressive neutralisation of the alkaline cement paste, and diffusional transport. The principal action of the external water is to dissolve the portlandite and decompose the remaining CSH gel; as a consequence the composition of the leachant solution will be affected by contact with cement. The scenarios considered in a safety analysis have to distinguish between two possible water conditions: static water or flowing water; the matrix decomposition (calcium release) will depend heavily on these conditions as well as on the chemical composition of the water. In a deep repository scenario of flowing water with a chemical composition containing HCO_3^- as the dominant species, the solubilization of calcium from $Ca(OH)_2$ and CSH will occur (Groves et al. 1991). In this situation, cement mass is chemically exposed to:

- The pH of the incoming water that produces a progressive neutralisation of the alkaline nature of the cement paste dissolving portlandite and CSH gel.
- In the presence of HCO_3^-, the release of calcium from $Ca(OH)_2$ and CSH is controlled by the precipitation of $CaCO_3$.
-

The general objectives of the work were to define the type of cement to be used in deep repositories analysing the role of the alkaline reserve when granitic water is used as leachant agent.

ACCELERATED TEST METHODS AND THEIR MECHANISM

Acid neutralisation test (ANT)

The progressive neutralisation of the alkaline cement paste may be followed by an acid attack and the measurement of the evolution of pH. It can be characterised by the milliequivalents of acid (H_3O^+) needed to get the particular pH value, per gram of sample.

The experimental procedure consisted of a batch test. Hydrated cement mixes are ground and sieved until a particle size less than 75 μm was obtained; then 10g of this solid (S) are mixed with deionized water (L) in a ratio S/L=1, producing an homogeneous slurry. Accelerated tests are performed in these slurries by adding an acid (HNO_3, 1N and aqueous CO_2, pH=4.1), at a rate of 0.5 $ml.min^{-1}$ with an automated titrator (Net Titrino 721 from Metrohm), and stirring the sample vigorously to favour continuous mixture of leachant and sample. In the case of nitric attack, a N_2 flow was maintained during the process to avoid carbonation. The evolution of pH was continuously recorded and liquid and solid phases were taken at different pHs corresponding to progressive degradation stages. The process of hydration was interrupted in each case by "freezing" the sample with acetone and ethanol; then samples were vacuum filtered through a 0.45μm filter. Further analysis of the leachates and solid phases were carried out according to the requirements for each special case; the experimental set-up is shown in Figure 1.

Effect of decreasing pH (protonation) simulated with HNO_3: Reactions between nitric acid and hydrated cement are based in the neutralisation of the cement alkalinity with formation of soluble calcium nitrate:

$$Ca(OH)_2 + 2HNO_3 \rightarrow Ca^{2+} + 2NO_3^- + 2H_2O$$

$$CaSiO_4H_2 \ (CSH) + 2HNO_3 \rightarrow Ca^{2+} + 2NO_3^- + SiO_2.2H_2O$$

For this reason, nitric acid can be used to simulate the pH decrease due to the dilution process that occur when groundwaters diffuse into the concrete. HNO_3 has the advantage of forming soluble salts which simplifies the study of acid attack.

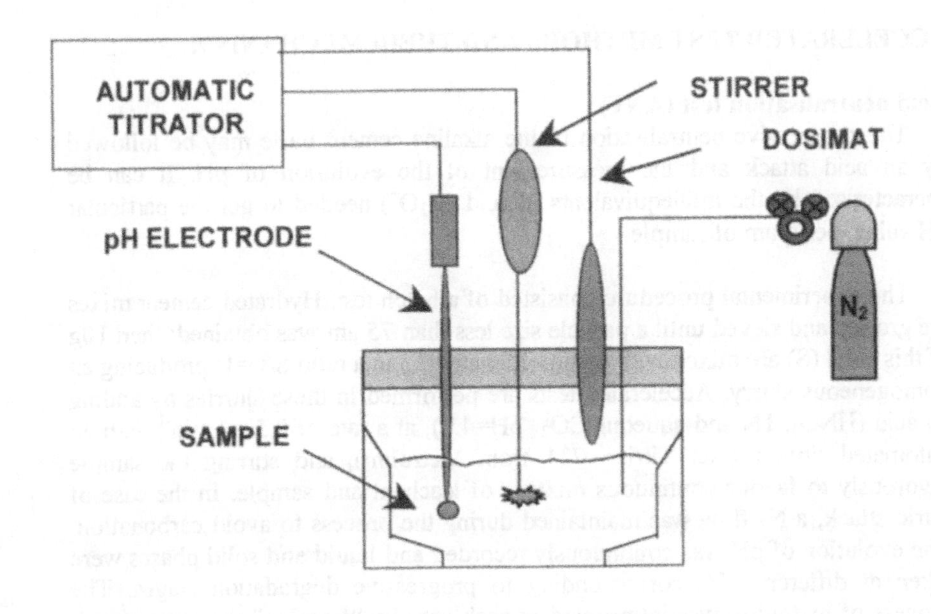

Figure 1. Experimental set-up

***Reactions in cement and concrete with CO_2 saturated water*: The use of carbonic acid is interesting due to its more realistic nature. Carbonic water is found to be very corrosive; there is a very strong dissolution of hydrated cement**

minerals and precipitation of large amounts of calcite (Cowie et al., 1991/92; Andac et al., 1999). The degradation process is due to a combination of:

1 Protonation and dilution by water (dissolution)
2 CO_2 (aq) attack (precipitation of $CaCO_3$).

Simplified reactions can be written as follows,

$$Ca(OH)_2 + H_2CO_3 \rightarrow CaCO_3 + 2H_2O$$

$$CaSiO_4H_2 \ \ (CSH) + H_2CO_3 \rightarrow CaCO_3 + SiO_2.2H_2O$$

Permeability test

Permeability experiments for different types of mortars in contact with (Ca,Mg)-bentonite, were carried out using granitic waters for periods of time not longer than 45 days. The objective of the work was to estimate the stability of concretes to be used for the construction of a HLW repository. Cylindrical mortar slices of 1 cm thick and bentonite samples of 1.5 cm thick, both of 5 cm diameter, were placed between two cylinders of metacrylate containing holes for water inlet and outlet. The block was sealed with an epoxy-resin in order to be sure that water pass only through the sample and measured fluxes are correct. Once the samples were placed in contact, a water head of 5 bars pressure was maintained to pass water from the concrete side to the bentonite after which, it was collected for analysis (see Fig. 2). The permeability of the samples and the pressure applied to the water regulate the water flow rate. The objective of this testing programme was to study the evolution of chemical and microstructural changes occurring when materials are subjected to a continuous flow of granitic water.

EXPERIMENTAL

Acid neutralisation test (ANT)

Materials: A Spanish cement (CEM I-SR according to the European Standard ENV 197-1) and two types of mineral additions (fly ash and blast furnace slag) were used for the testing.

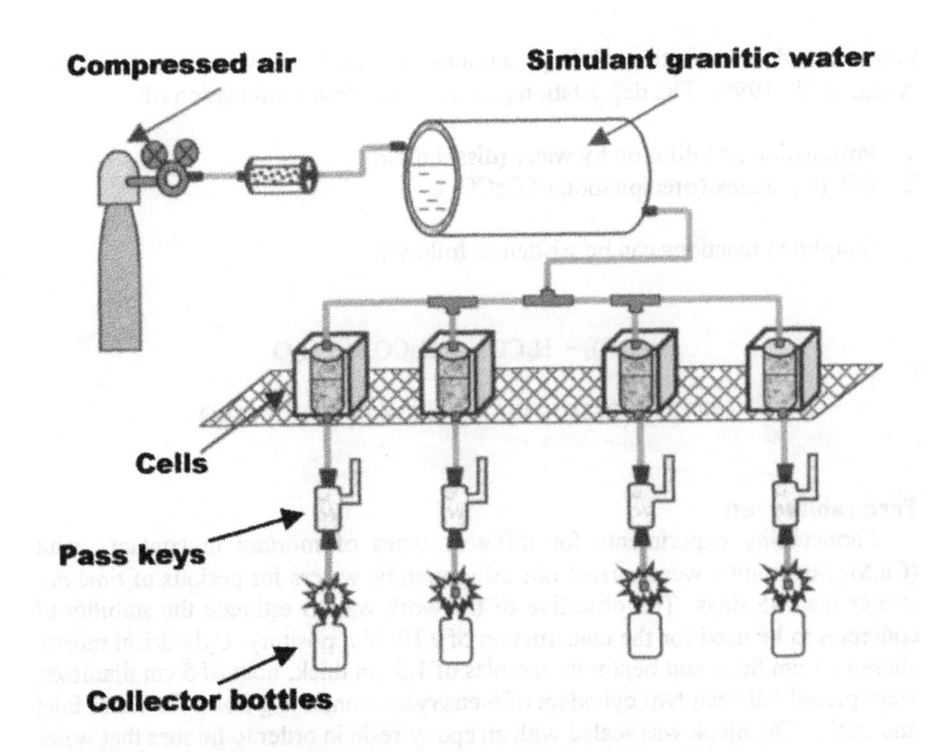

Figure 2. Experimental set-up for permeability tests

Chemical composition of materials: Chemical composition of mineral additions and cementitious mixtures used in this work is shown in Tables I and II respectively. Fly ash and slag were added to paste mixes on a cement replacement basis. A constant water-cement ratio (0.4) was chosen for pastes; then, when a supplementary cementing material is added, an equal volume of another component, in this case cement, has to be removed to keep the same comparison basis (Papadakis 2000). Cement pastes were hydrated in different ways: mixtures A, C, E, H, I, and J were hydrated 28 days in 98% relative humidity and 20±2°C, mixture G was hydrated in 98% relative humidity, 28 days at 20±2°C and 28 days at 40±0.1°C. Finally mixtures A, B, D and F were hydrated in 98% relative humidity for 28 days at 40±0.1°C.

Table I Chemical composition of mineral additions

CHEMICAL ANALYSIS (%)	SiO_2	Al_2O_3	Fe_2O_3	CaO (total)	MgO	SO_3	Na_2O	K_2O	S^{2-}
Fly Ash (FA)	45.25	25.20	9.25	6.50	1.90	0.20	0.46	3.32	-----
Slag (S)	35.78	11.81	0.33	41.20	7.97	-----	0.27	0.66	0.94

Table II Chemical composition of cements and mixes

CHEMICAL ANALYSIS (%)	Mix	SiO_2	Al_2O_3	Fe_2O_3	CaO (total)	MgO	SO_3	Na_2O	K_2O
100%Cem	A	19.90	2.86	4.40	65.04	2.13	3.70	0.11	0.45
90%Cem+10%FA	B	22.44	5.09	8.77	59.19	2.11	3.35	0.15	0.74
85%Cem+15%FA	C	23.70	6.21	5.13	56.26	2.10	3.18	0.16	0.88
80%Cem+20%FA	D	24.97	7.33	5.37	53.33	2.08	3.00	0.18	1.02
75%Cem+25%FA	E	26.24	8.45	5.61	50.41	2.07	2.83	0.20	1.18
70%Cem+30%FA	F	27.51	9.56	5.86	47.48	2.06	2.65	0.22	1.31
65%Cem+35%FA	G	28.77	10.68	6.10	44.51	2.05	2.48	0.23	1.45
85%Cem+15%S	H	22.28	4.20	3.70	61.46	3.01	3.47	0.13	0.48
75%Cem+25%S	I	23.87	5.10	3.38	59.08	3.59	3.32	0.15	0.50
50%Cem+25%FA +25%S	J	30.21	10.68	4.60	44.45	3.53	2.45	0.24	1.22

Permeability test

Three cells were prepared as described above; experimental conditions and materials used for experiments are as follows:

Experimental conditions:

- Temperature: 20±2°C .
- Hydraulic pressure: 0.5 MPa
- Length of experiments: ≤45 days.

Materials:

- Bentonite: (Ca,Mg)-bentonite 1.2g/cm^3 dry density. Mineralogical composition of the bulk sample showed a predominance of smectite with minor quantities of quartz, potassic feldspar, plagioclase, cristobalite, opal and calcite. The clay fraction was practically pure smectite.
- Mortars: 1:3, w/c=0.45, curing time: 10 days (concrete in the repository will be fresh).
- Cementitious materials (characterised above):

- CEM I-SR
- 85% CEM I-SR + 15% FA
- 85% CEM I-SR + 15% Slag.

- Groundwater: chemical composition of groundwater used in these experiments is shown in Table III.

Table III Chemical composition of granitic water

pH	[Ca^{2+}] mmol/l	[Na^{+}] mmol/l	[K^{+}] mmol/l	[$HCO3^{-}$] mmol/l	[Si^{4+}] mmol/l	[Mg^{2+}] mmol/l	[Cl^{-}] mmol/l	[SO_4^{2-}] mmol/l
7,91	0,840	0,551	0,024	2,510	0,351	0,277	0,274	0.18

Variables measured in every case were: flux and pH of the effluent. At the end of the experiment mortars were characterised by:

- X Ray diffraction.
- SEM observation and backscattered electron mode analysis.
- Mercury porosimetry.

XRD analysis of the starting unhydrous and hydrated materials, were performed using a Philips PW1710 equipment. Samples were observed and analysed using a JEOL JSM 5400 microscope, with EDX, from Oxford Instruments. For mercury intrusion porosimetry, a Micromeritics porosimeter model 9320, was used.

RESULTS

Acid Neutralisation Test

HNO_3 attack: Figure 3 presents the nitric acid degradation curves for mixes studied in this work. It can be seen that in the case of extreme degradation, blended cements show a higher acid degradation rate; that means that the higher amount of portlandite present in OPC cements and available for leaching results in a lower degree of apparent deterioration.

Experiments with CO_2 saturated water: Figure 4 presents the carbonic acid degradation curves for some mixes studied in this work. In this case blended cements show also a higher degradation rate.

Permeability tests

Figure 5 presents the permeability results for the different tested cementititous materials; calculated hydraulic conductivity coefficients (k in m/s) vs. time are represented in this figure.

Results from permeability in concretes or mortars, are complicated by the presence of cracks and other inhomogeneities. The permeability of concrete is, however, to a large extent controlled by the permeability of the cement paste, and this is controlled by capillary porosity.

Results indicate that mineral additions used by replacement of cement, increase the permeability of the mortar prepared with CEM I-SR cement, being the higher one the corresponding to the mixture including fly ash (having this mixture a lower CaO content and then, a lower $Ca(OH)_2$ content). Then, the reducing porosity effect when supplementary materials are used has low relevance when water is pressed through the concrete.

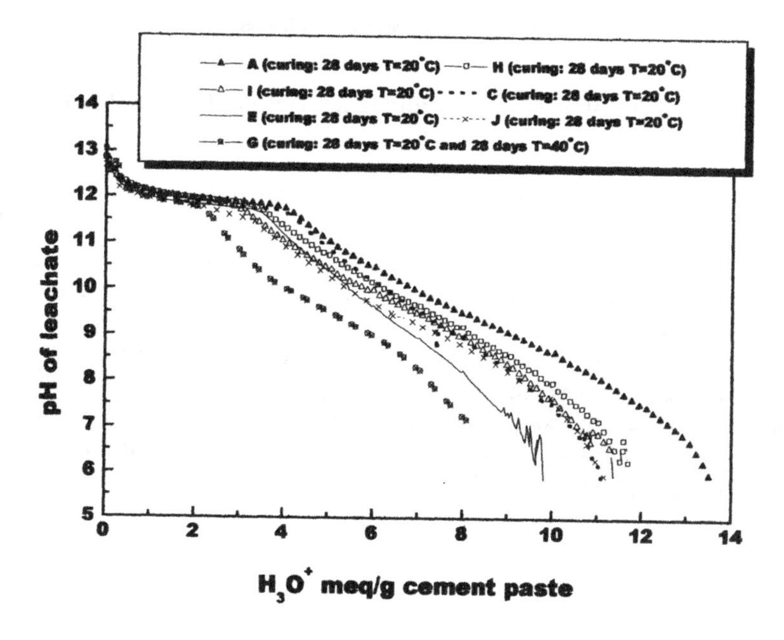

Figure 3. Nitric degradation curves for different mixes

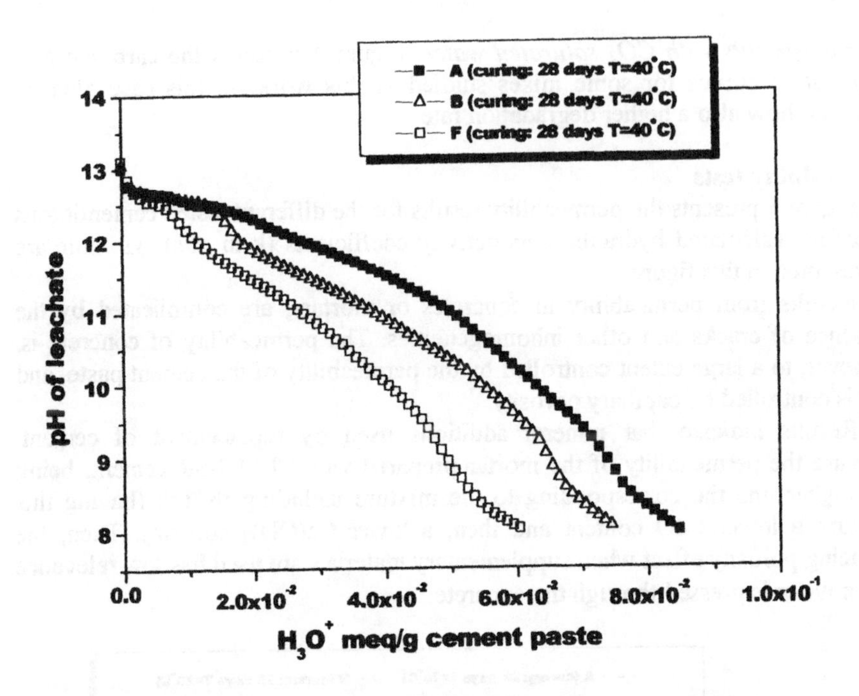

Figure 4. Degradation curves for different mixes in CO_2 saturated water

Mortar characterisation of samples from permeability tests:

X ray diffraction: Relative intensity of XRD spectra was calculated refered to C_4AF. Figure 6 shows the mineralogical evolution of calcite and portlandite in the degraded parts. Peaks of calcite decrease in intensity after the experiment and calcite was frequently observed in the interface with the bentonite by optical microscopy. However, in the case of the mortar prepared without mineral additions, portlandite peaks increase in intensity after the experiment; that means that in spite of a possible consumption of the existing portlandite by reaction with carbonates to form $CaCO_3$, new portlandite is precipitating; and the hypothesis suggested is the continued hydration.

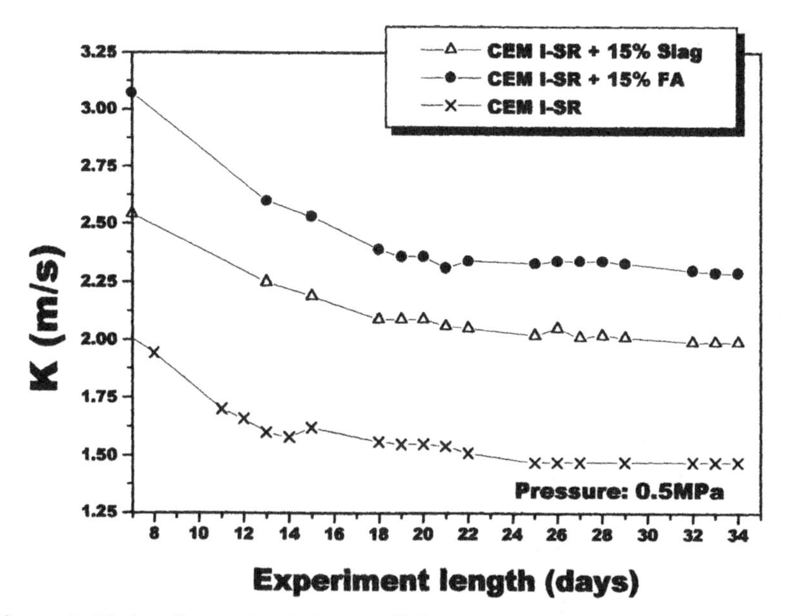

Figure 5. Hydraulic conductivity coefficients vs. time for different materials.

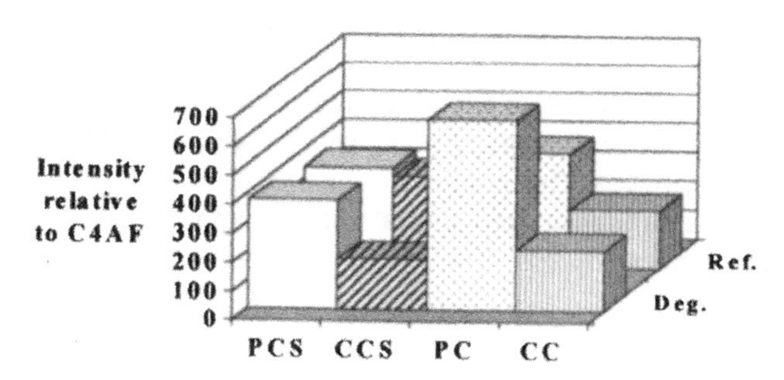

Figure 6. Mineralogical evolution of calcite and portlandite

Mercury Intrusion Porosimetry: Porosimetry measurements were made on average representative samples. It is known that, the pore volumes determined by the mercury porosimeter are not fully representative of the real pore volumes. The porosimeter only considers the pores as increasingly small cylinders whose diameters depend on the mercury pressure applied (Revertegat et al. 1997; Diamond 1999). If a high mercury pressure has to be applied to reach an isolated large pore, the pore will be recorded as a very small diameter pore (inkpot effect). Nevertheless, a correlation can be made between the change in the porosity of the sample and the predominant process causing a structural change (degradation, continued hydration).

The change in total porosity provides important information about the degradation process, but the degradation kinetics are connected with the change in the distribution of pore sizes. In this work, pre sizes have been arbitrarily subdivided into three main families:

- Pores>2μm (macrodefects and cracks).
- Pores between 0.05 and 2 μm (capillary pores).
- Pores<0.05μm (fine pores).
-

Figure 7 shows the variations in total porosity and in the three main pore families for reference sample, and degraded mortars.

In general, there are not significant changes in total porosity, except in the case of mortar prepared with fly ash, where total porosity shows a slight decrease after the test. This decrease could be explained as due to the pozzolanic reaction which could be progresively densifying the mortar and decreasing porosity. Notice however, that the lowest value of total porosity corresponds to the mortar prepared without mineral additions.

Pore size distribution shows that, in any case, the percentage of pores<0.05μm (fine pores) increases and the percentage of pores with a mean size between 0.05-2 μm (capillary pores) decreases. This implies that capillary pores are transformed into finer pores without a significant drop in total porosity, and it characterises the reorganization of the hydrates.

CFA: CEM I-SR+Fly Ash mortar
CS: CEM I-SR+Slag mortar
C: CEM I-SR mortar
Ref.: Non tested mortars
Deg.: Degraded mortars

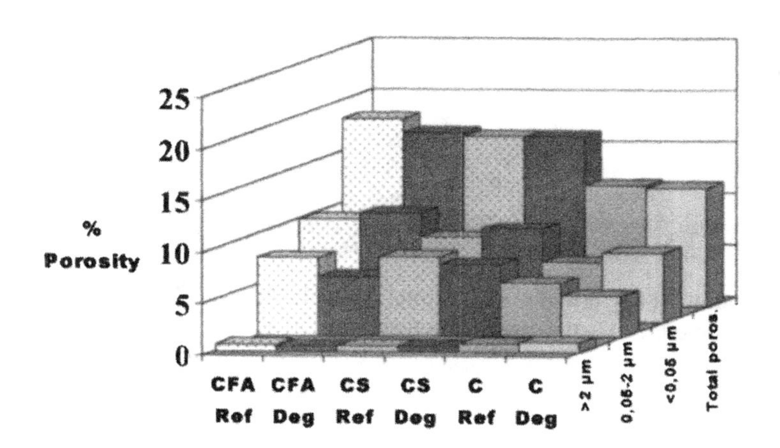

Figure 7. Porosity of reference samples and degraded mortars

SEM observation: A microscopic observation of the samples showed some erosion and decalcification at the mortar penetrating water surface. Hydration of anhydrous grains (Hadley grains) in the mortar without mineral additions could also be observed (see Figure 8).

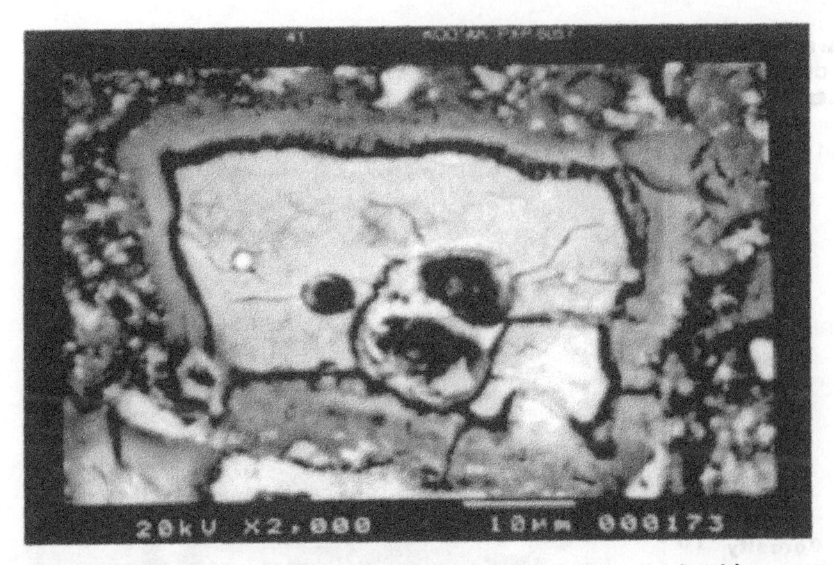

Fig. 8 Hydration of an initially anhydrous grain during the leaching process showing development of Hadley grain structures.

DISCUSSION

Acid neutralisation tests

Acid attack of cement paste has been described from many years ago (Hewlett, 1997), in spite of which, the trend of a particular sample may change depending on its composition, and on the intensity of the acid attack (defined by the acid $Ca(OH)_2$ dissolves during the acid attack and the pH value of the pore solution is maintained around 12.6 until all precipitated $Ca(OH)_2$ is removed. Then, the CSH starts to be decalcified and two scenarios may develop:concentration and the rate of contact with the paste sample).

It is known that the matrix $Ca(OH)_2$ acts as a buffer of the acid dissolution of the cement paste, that is,

1- The intensity of the acid attack is so that the decalcification rate is able to maintain the pH at levels above a value of 12 or,

2- The acid attack intensity is higher and pH drops.

For the same regime of acid attack, the length (amount of acid milliequivalents) of the buffering period (the plateau in the curves pH of leachate-acid milliequivalents at pH around 12) and the rate of pH decrease after this period contains an important information on the acid neutralisation ability of a particular cement paste. This is the aim of the here so called ANT.

In present results, cement mixtures with a higher CaO content were degraded slower than others with a lower CaO content; which has been interpreted that the presence of $Ca(OH)_2$ seems to be a guarantee of acid and in consequence water resistance.

Similar results were found when CO_2 saturated water was used; this seems to indicate that concretes to be used in nuclear repositories, should be prepared maintaining a high amount of carbonatable constituents : $Ca(OH)_2$ and CSH.

To better realise the results, figure 9 shows the acid equivalents needed to reach pH 12 and 11 for mixes A, C, E, G, H, I, J (see Table II) when attacked by HNO_3 acid. pH 12 was chosen because it has been assumed that at this value the total disappearing of portlandite has occurred; pH 11 was chosen because it is equally assumed that at this pH CSH gel has started to be destroyed. It is easy to notice that mixes with a higher content of CaO need a higher amount of acid to dissolve portlandite and start to attack CSH gel. Thus, mixes A, H and I consume a higher acid milliequivalents (a higher buffering ability, pH=12), but mix C may result more resistant to water degradation than mixes H and I, because it consumes more acid milliequivalents to reach pH 11 than those. Following this argument it can be deduced then, that CSH with a higher Ca/Si ratio is chemically more resistant to acid attack.

Permeability tests

Permeability tests where water is forced to pass through the pore network of mortars, is another kind of test to simulate groundwater attack to concrete in high level radioactive waste repositories. In this regime, the water hydraulic pressure may influence the results.

In present test, the interaction of the mortar with groundwater aims into the dissolution of coexisting primary solid phases, accompanied by precipitation of new ones (portlandite has been here identified, see figure 6).

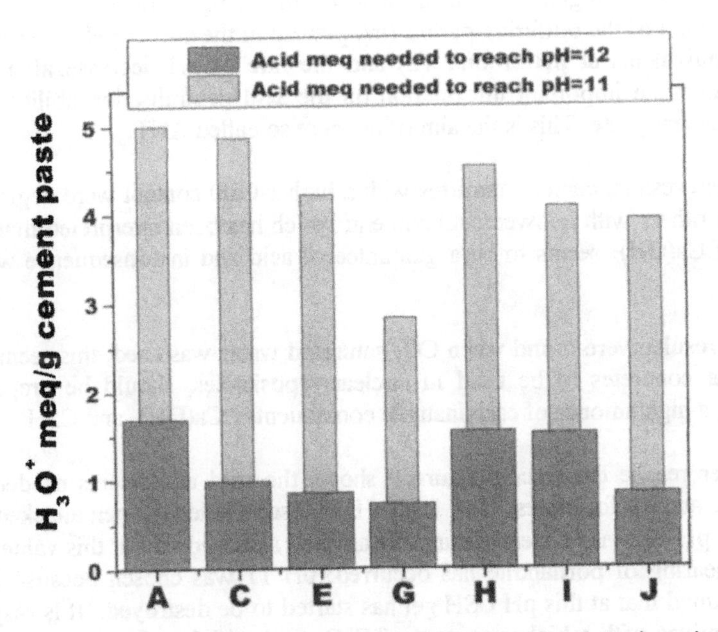

Figure 9. Acid equivalents needed to reach pH 12 and 11 for mixes A, C, E, G, H, I, J (Table 2) when attacked by HNO_3 acid

Regarding the results of porosimetry (see figure 7), the transformation of capillary pores in finer pores can be attributed to several factors including:

- The self-healing of the concrete: anhydrous grains are being hydrated, as could be observed by SEM, and new portlandite is being formed.
- The progress of the pozzolanic reactions.

In addition, the porosity analysis, together with XRD helps to corroborate the precipitation of secondary portlandite in the degradated CEM I-SR mortar. In consequence, the higher amount of CaO in the cement pastes and mortars, seems to have two beneficial effects:

- Chemical effect: the contact with water aims into a continued hydration of CaO producing secondary portlandite, which helps to enlarge the buffering capacity of the mix.

- Physical effect: the precipitation of secondary portlandite seems to occur in capillary pores, large pores are transformed into fine pores, densifying the paste.
-

XRD results also confirm decrease of calcium carbonate content inside the structure. The model proposed in the literature (Scherer 1999) indicates that calcite crystals could propagate through the pore network without resistance; then, the reason for the decrease in the intensity of calcite peaks observed by XRD could be the propagation of calcite through the pore network until the interface concrete-bentonite. In fact, calcite was observed in the interface with the bentonite by optical microscopy.

CONCLUSIONS

Two types of tests have been used to study the resistance of cement pastes and mortars to leaching by groundwater (granitic water) which also are informative to ascertain the role of the portlandite in the mechanisms of leaching by water, a) and acid neutralisation test (ANT) and b) a permeability test which forces the water to pass through the pore network.

From the results, the following conclusions have been drawn up:

- Both kind of tests have given interesting partial information on the overall process. They are then complementary.
- The dominant processes for the chemical interaction in the concrete/groundwater system, are the continued hydration with precipitation of secondary portlandite and CSH, and the dissolution of coexisting primary solid phases accompanied by precipitation and growth of secondary alteration phases.
- Precipitation of secondary portlandite has two beneficial effects:
  - Chemical: more precipitated $Ca(OH)_2$ enlarges the buffering capacity of the mix.
  - Physical: precipitation of secondary portlandite reduces porosity of mortars.

With respect to the role of cement replacement by a suplementary cementitious material:

- The total amount of CaO in the mix decreases; so the alkaline reserve is lower detected by a lower acid milliequivalents needed to reach a certain pH value.
- The reducing porosity effect when supplementary materials are used has low relevance when water is pressed through the concrete.

It can be then concluded as a preliminary deduction, in contrast to the opinion of other researchers, that concretes to be used in nuclear repositories should be based in having large amounts of primary or potential for secondary portlandite, to be more resistant to long term leaching.

REFERENCES

Adenot, F.; Buil, M.; "Modelling of the corrosion of the cement paste by deionized water", *Cement and Concrete Research*, **22**. 489-95 (1992).

Andac, M.; Glasser, F.P.; "Long-term leaching mechanisms of Portland cement-stabilized municipal solid waste fly ash in carbonated water", *Cement and Concrete Research*, **29** 179-86 (1999).

Cowie, J.; Glasser, F. P.; "The reaction between cement and natural waters containing dissolved carbon dioxide", *Advances in Cement Research*, **4**(15) 119-134 (1991/92).

Diamond, S.; "Aspects of concrete porosity revisited", *Cement and Concrete Research*, **29** 1181-88 (1999).

Faucon, P. Adenot, F; Jacquinot, J. F.; Petit, J. C.; Cabrillac, R; Jordá, M; "Long term behaviour of cement pastes used for nuclear waste disposal: review of physico-chemical mechanisms of water degradation", *Cement and Concrete Research*, **28** 847-57 (1998).

Groves, G. W.; Brough, A., Richardson, I. G.; Dolosou, C. M.; "Progressive changes in the structure of hardened C_3S cement pastes due to carbonation", *Journal of American Ceramic Society*, **74**[11] 2891-96 (1991).

Hewlett, P. C.; "Lea's chemistry of cement and concrete", John Wiley & Sons, 1997.

Papadakis, V. G.; "Effect of supplementary cementing materials on concrete resistance against carbonation and chloride ingress", *Cement and Concrete Research*, **30** 291-99 (2000).

Revertegat, E.; Adenot, F.; Richet, C.; Wu, L.; Glasser, F. P.; Damidot, D.; Stronach, S. A.; "Theoretical and experimental study of degradation mechanisms of cement in the repository environment", CEC Contract nº FI2W-CT90-0035, Final Report. ISBN 92-828-0394-5, 1997.

Scherer, G. W.; "Crystallization in pores", *Cement and Concrete Research*, **29** 1347-1358 (1999).

Unsworth, H.P.; Lota, J.S.; "Microstructural and chemical changes during leaching of cementititous materials"; 4 pp. in *Performance and Durability of Cementitious Materials*, **4**. Edited by H. Justnes. Amarkai AB and Congrex Göteborg AB, Göteborg (Sweden), 1997.

Van der Sloot, H.A.; "Leaching behaviour of waste and stabilised waste materials; characterization for environmental assessment purposes", *Waste Management & Research*, **8** 215-228 (1990).

Anik Delagrave and Felek Jachimowicz

David Myers and Michael Thomas

INFLUENCE OF CALCIUM HYDROXIDE DISSOLUTION ON THE TRANSPORT PROPERTIES OF HYDRATED CEMENT SYSTEMS

Jacques Marchand[1,2], Dale Bentz[3], Eric Samson[1,2] and Yannick Maltais[1,2]

[1]CRIB - Department of Civil Engineering,
Laval University, Canada, G1K 7P4

[2]SIMCO Technologies Inc.,
1400 boul. du Parc Technologique, Québec, Canada, G1P 4R7

[3]Building and Fire Research Laboratory
National Institute of Standards and Technology, Gaithersburg, MD 20899, USA

ABSTRACT

Calcium hydroxide is one of the main reaction products resulting from the hydration of Portland cement with water. It is also one of the more soluble phases found in hydrated cement systems. The influence of calcium hydroxide dissolution and its effect on the diffusion properties of hydrated cement pastes were investigated using the NIST CEMHYD3D cement hydration and microstructure development model. The results of these simulations were implemented in another numerical model called STADIUM. This latter model can be used to predict the transport of ions in unsaturated porous systems. Numerical simulations clearly indicate that calcium hydroxide dissolution contributes to a marked increase in the porosity of the hydrated cement paste. This increase in porosity has a detrimental influence on the material transport properties. The results yielded by the numerical simulations are in good agreement with data of calcium leaching experiments performed in deionized water.

INTRODUCTION

Calcium hydroxide (CH), along with *C–S–H*, are the end products of the reaction of alite and belite with water. The abundance of CH in the hydrated cement paste varies with the degree of hydration of the cement, and can reach approximately 26% of the total volume of a mature paste. Contrary to the *C–S–H* gel that is an ill-crystallized phase, CH is present predominantly in the form of well-defined crystals in the hydrated cement paste. The size of these crystals tends to vary significantly from one to approximately 100 microns in diameter[1,2].

The leaching of calcium may be a matter of concern for the durability of concrete[3]. In some cases, CH dissolution and the decalcification of C–S–H may increase the porosity of the surface layers of concrete, and detrimentally affect the resistance of the material to deicer salt scaling and ion penetration[4]. In other instances, the leaching of calcium may also affect the core of the material and have a negative influence on the engineering properties of concrete structures[3]. For instance, CH dissolution and C–S–H decalcification have been found to have a detrimental influence on the mechanical and transport properties of hydrated cement systems[5–9].

Over the years, several studies have clearly demonstrated that the investigation of calcium leaching mechanisms by laboratory experiments is often difficult and generally time-consuming[8,9]. Furthermore, since both phenomena readily affect the pore structure of the material, the kinetics of CH dissolution and C–S–H decalcification quickly become non-linear, and a reliable prediction of the evolution of the concrete properties upon leaching can hardly be made on the sole basis of experimental results.

This paper presents the main results of a study of the influence of CH dissolution on the mechanisms of ion transport in hydrated cement systems. The effects of CH dissolution on the pore structure and transport properties of various hydrated cement paste mixtures were investigated using the NIST CEMHYD3D cement hydration and microstructure development model[10,11]. These results were then implemented in another numerical model called STADIUM[12,13]. This model can be used to predict the transport of ions and water in reactive porous materials (such as concrete). The degradation characteristics of neat cement paste mixtures immersed for 3 months in deionized water served as a basis for the validation of the model.

COMPUTER MODELING OF MICROSTRUCTURE AND DEGRADATION

Over the past decade, the development of numerical models has provided new tools to investigate the influence of calcium leaching on the evolution of the properties of cement-based materials. A few years ago, Bentz and Garboczi[14] used a cellular automaton-type digital-image-based model to study the influence of CH dissolution on the pore structure of hydrated C_3S pastes. They showed that leaching mechanisms can have a detrimental effect on the connectivity of the pore structure of the material.

Table 1: Chemical and mineralogical properties of the three cements

Oxides	Cement		
	A: CSA Type 10	B: CSA Type 50	C: White
SiO_2	19.78	21.45	24.29
Al_2O_3	4.39	3.58	1.71
TiO_2	0.22	0.21	0.07
Fe_2O_3	3.00	4.38	0.32
CaO	62.04	63.93	68.60
SrO	0.26	0.07	0.13
MgO	2.84	1.81	0.54
Mn_2O_3	0.04	0.05	0.03
Na_2O	0.32	0.24	0.14
K_2O	0.91	0.70	0.03
SO_3	3.20	2.28	2.11
LOI	2.41	0.86	1.13
Bogue			
C_3S	59	62	77
C_2S	12	16	12
C_3A	7	2	4
C_4AF	9	13	1

The cement hydration and microstructure development model was later modified to account for the physical and mineralogical characteristics of cement grains on the properties of hydrated systems[10,11]. This new model, called CEMHYD3D, was used to investigate the effects of CH dissolution on the pore structure and diffusion properties of two series of hydrated cement pastes prepared at w/c ratios of 0.4 and 0.6, respectively. Hydrated microstructures were created by considering the characteristics of three commercial cements (A: CSA Type 10, B: CSA Type 50 and C: a Danish white cement). The chemical and mineralogical properties of these cements are given in Table (1).

The starting three dimensional microstructures were based on the measured particle size distributions of the cement powders and two-dimensional SEM/X-ray image sets in which the clinker phases had been individually identified[15]. The starting microstructures were then hydrated either for 2000 cycles or until achieving a degree of hydration, α, that corresponded to the experimentally measured value (based on non-evaporable water measurements). The diffusivities of these "final" microstructures were then computed using the techniques described in the next section. These final microstructures were used as input microstructures for the leaching program. The CH in the microstructures was progressively leached as described previously[14], and the diffusivity of the leached microstructures determined. In this way, the relative increase in diffusivity due to the leaching of CH from a hydrated microstructure could be assessed.

COMPUTATION OF DIFFUSIVITY

An electrical analogy is used to compute the relative diffusivity of the composite media[16], where relative diffusivity is the ratio of the diffusivity of an ion in the composite media relative to its value in bulk water (proportional to the inverse of the formation factor). Conductivities are assigned to each phase comprising the microstructure and the resultant composite relative conductivity is computed[17] and related to a relative diffusivity using the Nernst-Einstein relation[16]:

$$\frac{D}{D_0} = \frac{\sigma}{\sigma_0} \tag{1}$$

where σ/σ_0 is the computed relative conductivity and D/D_0 is the relative diffusivity for the random microstructure. For this study, capillary pores (ϕ) are assigned a relative diffusivity of 1.0, while the $C\text{–}S\text{–}H$ gel is assigned a relative diffusivity of 0.0025 (1/400)[17].

Previous studies have indicated that diffusivities computed using these relative values compare favorably to those measured experimentally[17–18]. For each hydrated/leached microstructure, the diffusivity was computed in each of the three principal directions and the average value reported.

MODELING IONIC TRANSPORT IN CEMENT SYSTEMS

As previously mentioned, the numerical results yielded by the NIST CEMHYD3D model were implemented in another model called STADIUM[12,13]. This latter model has been developed to predict the transport of ions in unsaturated porous media. The model also accounts for the effect of dissolution/precipitation reactions on the transport mechanisms.

The description of the various transport mechanisms relies on the homogenization technique. This approach first requires writing all the basic equations at the microscopic level. These equations are then averaged over a Representative Elementary Volume (REV) in order to describe the transport mechanisms at the macroscopic scale[19,20].

In this model, ions are considered to be either free to move in the liquid phase or bound to the solid phase. The transport of ions in the liquid phase at the microscopic level is described by the extended Nernst-Planck equation[21] to which is added an advection term. After integrating this equation over the REV, the transport equation becomes:

$$\frac{\partial((1-\phi)C_{is})}{\partial t}+\frac{\partial(\theta C_i)}{\partial t}-\frac{\partial}{\partial x}\left(\theta D_i\frac{\partial C_i}{\partial x}+\theta\frac{D_i z_i F}{RT}C_i\frac{\partial\Psi}{\partial x}+\theta D_i C_i\frac{\partial \ln\gamma_i}{\partial x}-C_i V_x\right)=0 \quad (2)$$

where the uppercase symbols represent the variables averaged over the REV. In equation (2), C_i is the concentration of the species i in the aqueous phase, C_{is} is the concentration in solid phase, θ is the volumetric water content (expressed in m^3/m^3 of material), D_i is the diffusion coefficient, z_i is the valence number of the species, F is the Faraday constant, R is the ideal gas constant, T is the temperature of the liquid, Ψ is the electrical potential, γ_i is the chemical activity coefficient and V_x is the velocity of the fluid. Equation (2) has to be written for each ionic species present in the system.

To calculate the chemical activity coefficients, several approaches are available. However, models such as those proposed by Debye-Hückel or Davies are unable to reliably describe the thermodynamic behavior of highly concentrated electrolytes such as the hydrated cement paste pore solution. A modification of the Davies equation described in reference[22] was found to yield good results.

The Poisson equation is added to the model to evaluate the electrical potential Ψ. It relates the electrical potential to the concentration of each ionic species[23]. The equation is given here in its averaged form:

$$\frac{\partial}{\partial x}\left(\theta\tau\frac{\partial\Psi}{\partial x}\right)+\theta\frac{F}{\epsilon}\sum_{i=1}^{N} z_i C_i = 0 \quad (3)$$

where N is the total number of ionic species, ϵ is the dielectric permittivity of the medium, in this case water, and τ is the tortuosity of the porous network.

The velocity of the fluid, appearing in equation (2) as V_x, can be described by a diffusion equation when its origin is in capillary forces present during drying/wetting cycles[24]:

$$V_x = -D_w\frac{\partial\theta}{\partial x} \quad (4)$$

where D_w is the non-linear water diffusion coefficient. This parameter varies according to the water content of the material[24].

To complete the model, the mass conservation of the liquid phase must be taken into account[24]:

$$\frac{\partial\theta}{\partial t}-\frac{\partial}{\partial x}\left(D_w\frac{\partial\theta}{\partial x}\right)=0 \quad (5)$$

As can be seen, moisture transport is described in terms of a variation of the (liquid) water content of the material. It should be emphasized that the choice of using the material water content as the state variable for the description of this problem has an important implication on the treatment of the boundary conditions. Since the latter are usually expressed in terms of relative humidity, a conversion has to be made. This can be done using an adsorption/desorption isotherm[24].

The first term on the left-hand side of equation (2) (in which C_{is} appears), accounts for the ionic exchange between the solution and the solid[19]. It can be used to model the influence of precipitation/dissolution reactions on the transport process. More information on this procedure can be found in reference[12].

The transport of ions and water in unsaturated cement systems can be fully described on the basis of equations (2) to (5). Previous experience[12] has shown that most practical problems can be reliably described by seven different ionic species (OH^-, Na^+, K^+, SO_4^{2-}, Ca^{2+}, $Al(OH)_4^-$ and Cl^-) and five solid phases (CH, *C-S-H*, ettringite, gypsum and hydrogarnet).

The input data required to run the model can be easily obtained. The initial composition of the material (i.e. its initial content in CH, ettringite, ...) can be easily calculated by considering the chemical (and mineralogical) make-up of the binder, the characteristics of the mixture and the degree of hydration of the system[25].

The model also requires determining the initial composition of the pore solution and the porosity of the material. Samples of the pore solution of most hydrated cement systems can be obtained by extraction using the technique described by Longuet et al.[26]. The total porosity of the material can easily be determined in the laboratory following standardized procedures (such as ASTM C642)[27].

Some information on the transport properties of the material is also required to run the model. The ionic diffusion properties of the solid can be determined by a migration test[12]. The water diffusion coefficient of the material can be assessed by nuclear magnetic resonance imaging[24].

LEACHING EXPERIMENTS

In order to validate the results of the numerical simulations, six different cement paste mixtures were prepared with the three cements described in Table (1) and at two water/cement ratios (0.4 and 0.6). Only the results obtained for the 0.6 water/cement ratio mixture made of the CSA Type 10 cement (cement A) will be reported. Data obtained for the other mixtures will be discussed in a forthcoming publication.

All mixtures were prepared using deionized water. The mixtures were batched in a high-speed mixer placed under vacuum (at 10 mbar) to prevent, as much as possible, the formation of air voids during mixing. Mixtures were cast in plastic cylinders (diameter = 7.0 cm, height = 20 cm). The molds were sealed and rotated for the first 24 hours in order to prevent any bleeding of the mixtures. At the end of this period, the cylinders were demoulded and sealed with an adhesive aluminum foil for an 18-month period at room temperature. This period was selected in order to get mature and well-hydrated cement pastes.

After the 18-month curing period, samples of each mixture were subjected to migration tests, porosity measurements pore solution extractions and thermal analyses (to assess the degree of hydration of each system). The experimental procedures for the migration tests and the pore solution extractions have been described elsewhere[28,29]. Porosity measurements were carried out according to the requirements of ASTM C 642[27]. The water diffusion properties of these mixtures had been previously determined by Nuclear Magnetic Resonance Imaging (NMRI) as part of a previous project[30].

The remaining parts of the cement paste cylinders were sawn in thin disks. The thickness of the disks varied from 12-15 mm. The disks were then vacuum saturated in a sodium hydroxyde solution (prepared at 300 mmol/l) for a 24-hour period prior to the degradation experiments. The latter were performed during three months under saturated (series 1) and unsaturated conditions (series 2) using deionized water.

The series 1 samples were first coated with an epoxy resin (on all their faces except one) and then immersed in the test solutions (see Figure (1)). For the samples of series 2 (unsaturated conditions), a relative humidity gradient was created between the two faces of the disks (see Figure (1)). One face was directly placed in contact with water and the other was placed in contact with a CO_2 free environment at a relative humidity close to 65%. In order to avoid carbonation, nitrogen was added on a daily basis in the compartment.

At the end of the degradation experiments, samples were broken in small pieces and then immersed in isopropyl (propan-2-ol) alcohol for a minimum period of 14 days. After this period, samples were dried under vacuum for 7 days. Once the drying process was completed, the samples were impregnated with an epoxy resin, polished, and coated with carbon.

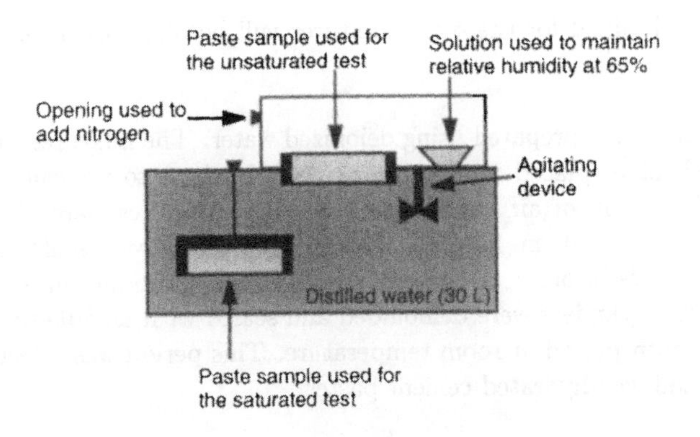

Figure 1: Degradation test set-up.

Microstructural alterations of the cement paste samples were investigated by means of electron microprobe analyses. The polished sections were observed using a microprobe (Cameca SX-100) [1] operating at 15 kV and 20 nA. For each sample, measurements were performed at a maximum interval of 13 microns on four distinct imaginary lines extending from the external surface in contact with the aggressive solution toward the internal part of the samples. At each point of measurement, the total content of calcium, sulfur, sodium, potassium, silicon, and aluminum was determined.

RESULTS AND DISCUSSION

Numerical results obtained with the NIST CEMHYD3D cement hydration and microstructure development model clearly indicate that leaching has a significant effect on the diffusivity of cement pastes. The diffusivity values calculated with the model for the original and completely leached microstructures are summarized in Table (2). In agreement with previous results for simpler C_3S systems[14], the increase in diffusivity due to leaching is seen to be a factor of 20 or more, depending on the initial w/c ratio and the degree of hydration achieved prior to leaching. Additionally, the increase in diffusivity is seen to be much more dramatic for the lower w/c ratio systems, due mainly to the re-percolation of the capillary pore network during the leaching of the CH. For the higher w/c ratio systems, the depercolation of the

[1] Certain commercial equipment is identified by name in this paper to adequately specify the experimental procedure. In no case does such identification imply endorsement by the National Institute of Standards and Technology, nor does it imply that the products are necessarily the best available for the purpose.

capillary porosity is never achieved during the initial hydration (since the critical percolation threshold for the capillary porosity is on the order of 0.20)[14,17,18]. Thus, the relative increase in diffusivity caused by leaching is significantly less since both the hydrated and hydrated/leached microstructures contain a continuous capillary pore system.

Table 2: Model results for increase in diffusivity of cement pastes due to CH leaching

Cement	w/c	α	ϕ_{orig}	$(D/D_0)_{orig}$	ϕ_{leach}	$(D/D_0)_{leach}$	D_l/D_o
A	0.40	0.6823	0.216133	0.00465	0.365127	0.0726	15.62
A	0.40	0.7985	0.164761	0.00222	0.333371	0.0535	24.1
A	0.60	0.7234	0.383334	0.0647	0.502002	0.1877	2.90
A	0.60	0.8879	0.328439	0.0322	0.467383	0.1506	4.68
B	0.40	0.7141	0.200007	0.00338	0.346106	0.0569	16.86
B	0.40	0.8101	0.159915	0.00204	0.320229	0.0423	20.77
B	0.60	0.790	0.342954	0.0377	0.466465	0.141	3.74
B	0.60	0.9010	0.308901	0.0223	0.444681	0.1207	5.41
C	0.40	0.7364	0.188660	0.00286	0.353968	0.0617	21.57
C	0.40	0.8142	0.153031	0.0022	0.330547	0.0493	22.4
C	0.60	0.7583	0.356682	0.0438	0.489645	0.1689	3.86
C	0.60	0.8917	0.307971	0.0222	0.456563	0.1366	6.15

In addition to leaching all of the CH from a hydrated microstructure, a fraction of the CH can be leached by specifying a leaching probability and a number of leaching cycles to be executed in the leaching program. For each leaching cycle, the microstructure is first scanned to identify all CH pixels which are in contact with capillary porosity. In a second scan, these pixels are randomly leached in proportion to the user-specified leaching probability. For six of the microstructures summarized in Table (2), "partial" leaching of the microstructures has been executed. The results for the ratio of the diffusivity of the leached to that of the original hydrated microstructure as a function of the amount of CH leached are provided in Figure 2. As observed previously[14] and in agreement with the available experimental data[30], initially the removal of a small portion of the CH due to leaching has only minor effects on the computed diffusivities. Then, as 30% to 60% of the CH is leached, the effects on diffusivity are more dramatic. Finally, above 90% leached, the increase in diffusivity tends to level off once more. Once again, it is clearly observed that the relative increase in diffusivity due to leaching is significantly greater for the lower 0.4 w/c ratio systems. Conversely, the differences between the three different cements at a constant w/c ratio are relatively minor, especially for CH leached fractions below about 50%.

In order to develop a simple equation for predicting the diffusivity ratio as a function of the fraction of CH leached from a microstructure, all of the data in Figure 2 were normalized using the diffusivity ratios for the original, $(DR(CH = 0))$, and

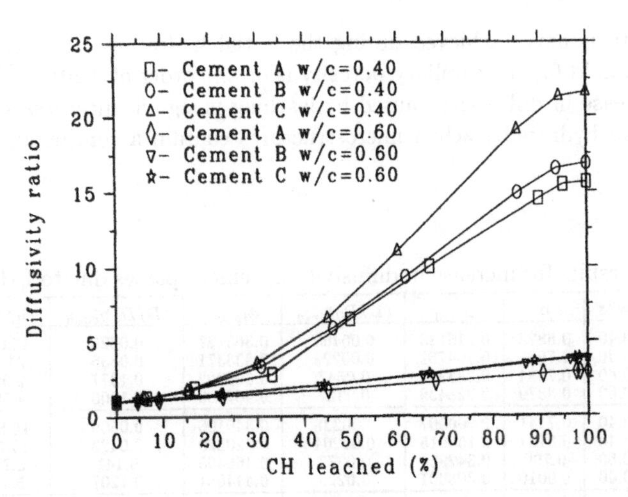

Figure 2: Model results for increase in diffusivity of cement pastes due to leaching of CH

completely leached, $(DR(CH = 100))$, microstructures and the following equation:

$$D_N(CH) = 1 + \frac{DR(CH) - DR(CH = 0)}{DR(CH = 100) - DR(CH = 0)} \tag{6}$$

It can be easily observed that this will result in normalized diffusivities (D_N) with values between 1 and 2 for every case.

These normalized diffusivity ratios are plotted vs. the fraction of CH leached for each of the six cement paste systems in Figure (3). While there is still some scatter amongst the different systems, particularly for small values of the CH leached, for engineering purposes, all of the data in Figure (3) has been fitted to the following equation:

$$D_N(CH) = 1 + \frac{1.1 \times CH^2}{0.28 + 0.79 \times CH} \tag{7}$$

where CH is the fraction of CH leached, having values in the range of [0,1]. As indicated by the dashed line in Figure (3), this equation gives an acceptable "average" fit to all of the data and will provide a simple method for estimating the diffusivity ratio for intermediate CH leaching when the values for the original and completely leached microstructures have been measured or computed.

Also included in Figure (3) are the upper and lower bounds computed using both the series/parallel and Hashin-Shtrikman equations[31]. In this case, it is assumed

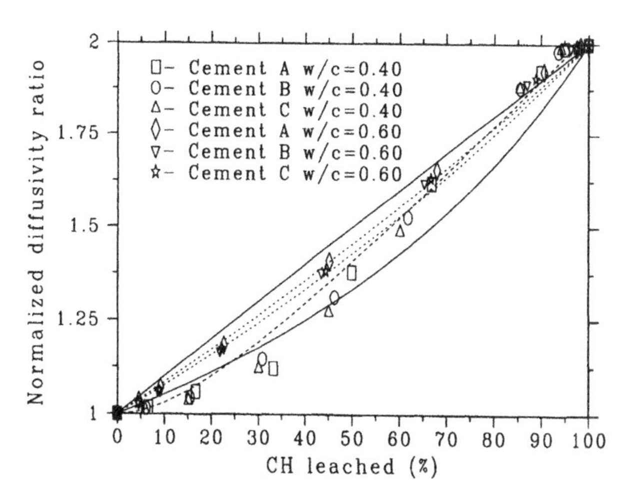

Figure 3: Model results for normalized diffusivity ratio as a function of amount of CH leached from the microstructure. Solid lines indicate series/parallel bounds for a simple two-phase composite, dotted lines are the Hashin-Shtrikman upper and lower bounds, and dashed line is the fit of equation 7 to all of the computer simulation data.

that the partially leached microstructure is composed of two components: original unleached cement paste with a normalized diffusivity of 1 and totally leached cement paste with a normalized diffusivity of 2. The fact that many of the plotted data points lie outside of these bounds (particularly the more restrictive Hashin-Shtrikman bounds) indicates that the simple consideration of a partially leached microstructure as a composite of unleached and totally leached phases is not totally appropriate but still serves as a useful abstraction.

The data provided by the NIST model were implemented in STADIUM and simulations were run for the 0.6 water/cement ratio mixture made of the CSA Type 10 cement (cement A) tested in saturated and unsaturated conditions. All the input data used in the simulations are summarized in Tables (3) and (4).

Results of the simulations are given in Figures (4) and (5). The total concentrations in calcium (expressed in g/kg of paste) yielded by the model are compared to the experimental calcium profiles provided by the microprobe analyses. The curve labeled "with damage factor" corresponds to the results calculated by taking into account the effect of CH dissolution on the transport properties of the paste mixtures (see Table (2) and Figure (2)). Microprobe data are given in counts per second (Cps).

Table 3: Physical characteristics of the two mixtures

Mixture	Diffusion coefficient (m^2/s)		Porosity (%)	α (%)
W/C=0.40	OH^-	7.6e-11	37	68
	Na^+	1.9e-11		
	SO_4^{2-}	2.2e-11		
	K^+	2.8e-11		
	Ca^{2+}	1.1e-11		
W/C=0.60	OH^-	17.6e-11	52	75
	Na^+	4.5e-11		
	SO_4^{2-}	5.2e-11		
	K^+	6.5e-11		
	Ca^{2+}	2.6e-11		

Table 4: Pore solution chemistry

Ion	Concentration (mmol/L) W/C = 0.40	W/C = 0.60
OH^-	700	434
Na^+	192	111
SO_4^{2-}	44	4
K^+	592	327
Ca^{2+}	2	2

As can be seen, whatever the moisture state of the samples, the calcium profiles predicted by the model are in good agreement with the profiles obtained by the microprobe analyses. Not only does the model accurately predict the depth of CH penetration, but it also reproduces quite well the total distribution in calcium over the entire thickness of the samples. It should be emphasized that the model has no "fitting parameter", and that the numerical simulations are solely based on the properties of the mixture and the chemical damage equation derived from numerical data provided by the NIST model. However, since the model predicts an averaged concentration per unit volume (or unit mass) of material, the numerical results do not reproduce the local variations in calcium measured by the microprobe analyses. These variations are due to the experimental "noise" of the technique and the presence of CH crystals within the hydrated cement paste matrix.

Both series of results also indicate that the increase in diffusivity induced by the removal of CH has a limited influence on the kinetics of degradation. This phenomenon can be explained by the fact that the duration of the experiments was

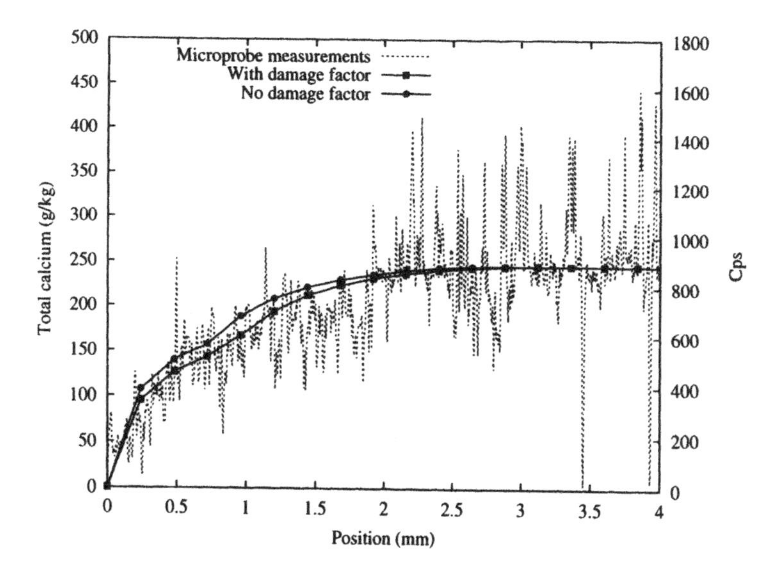

Figure 4: Total calcium profile after three months of immersion - CSA Type 10, W/C = 0.6, saturated conditions

limited to only 3 months. As can be seen, degradation is limited to the first two millimeters near the surface. For longer exposure periods (and thicker samples), the influence of CH dissolution would most certainly has a more significant effect on the degradation kinetics. Simulations ran for a 100-mm thick sample clearly indicate that chemical damage (i.e. the microstructural alterations induced by the dissolution of CH) does have a strong influence on the kinetics of penetration of the degradation front.

It should also be emphasized that experiments were carried out using a relatively porous mixture (prepared at a water/cement ratio of 0.6) which is less likely to be affected by the dissolution of CH. As previously mentioned, the pore structure of the 0.6 water/cement ratio mixture was already percolated, thus reducing the detrimental influence of CH dissolution on the transport properties of the material.

Additional simulations were run for the 0.4 water/cement ratio mixture prepared with the CSA Type 10 cement (cement A). Numerical results are given in Figure (6). Although the relative increase in diffusivity caused by leaching was found to be significantly more important for the 0.4 water/cement ratio mixtures, both series of results are similar. This phenomenon can be explained by the fact that the initial diffusion coefficient of the 0.4 water/cement ratio mixture was low,

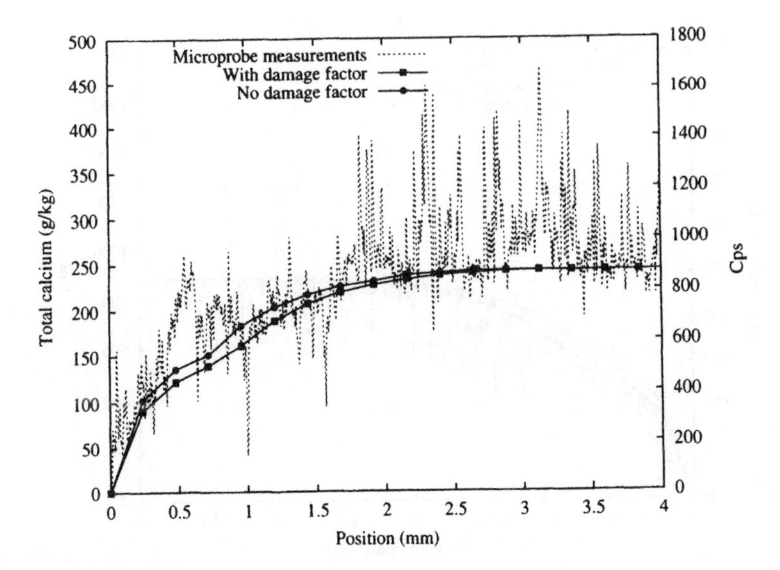

Figure 5: Total calcium profile after three months of exposure - CSA Type 10, W/C = 0.6, unsaturated conditions

thus reducing the rate of diffusion of calcium and hydroxide ions out of the sample. As can be seen in the figure, degradation is mainly limited to the first millimeter near the surface. These results clearly emphasize the importance of the diffusion coefficient of the material on the kinetics of degradation.

CONCLUDING REMARKS

The dissolution of CH was found to have a significant effect on the diffusivity of hydrated cement pastes. The increase in diffusivity due to leaching is seen to be a factor of 20 or more, depending on the initial w/c ratio and the degree of hydration achieved prior to leaching.

The increase in diffusivity is seen to be much more dramatic for the lower w/c ratio systems, due mainly to the re-percolation of the capillary pore network during the leaching of the CH.

Degradation profiles computed using the chemical damage equation provided by the NIST model and STADIUM compare favorably to experimental values.

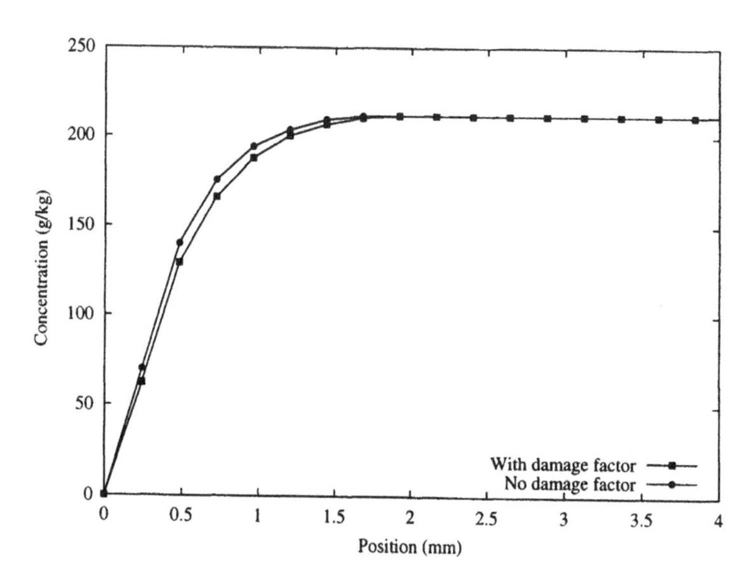

Figure 6: Total calcium profile after three months of immersion - CSA Type 10 W/C = 0.4

ACKNOWLEDGEMENTS

The authors are grateful to the Natural Sciences and Engineering Research Council of Canada and to SIMCO Technologies inc. for their financial support for this project.

REFERENCES

[1] St-John, D.A., Poole, A.W., Sims, I. (1998), *Concrete petrography - A handbook of investigative techniques*, Arnold , London, UK.

[2] Berger, R.L., McGregor, J.D. (1972), *Influence of admixtures on the morphology of calcium hydroxide formed during tricalcium silicate hydration*, Cement and Concrete Research, Vol. 2, pp. 43-55.

[3] Marchand, J., Beaudoin, J.J., Pigeon, M. (1999), *Influence of $Ca(OH)_2$ dissolution on the properties of cement systems*, Materials Science of Concrete - Sulfate Attack Mechanisms, American Ceramic Society, pp. 283-293.

[4] Marchand, J., Pleau, R., Gagne, R. (1995), *Deterioration of concrete due to freezing and thawing*, Materials Science of Concrete, Vol. IV, American Ceramic Society, pp. 283-354.

[5] Terzaghi, R.D. (1948), *Concrete deterioration in a shipway*, Journal of the

American Concrete Institute, Vol. 44, No. 6, pp. 977-1005.

[6] Tremper, B. (1931), *The effects of acid waters on concrete*, Journal of the American Concrete Institute, Vol. 28, No. 9, pp. 1-32.

[7] Carde, C., François, R. (1997), *Effect of the leaching of calcium hydroxide from cement paste on the mechanical and physical properties*, Cement and Concrete Research, Vol. 27, pp. 539-550.

[8] Adenot, F., Buil, M. (1992), *Modelling the corrosion of the cement paste by deionized water*, Cement and Concrete Research, Vol. 22, pp. 489-496.

[9] Delagrave, A., Gerard, G., Marchand, J. (1997), *Modelling calcium leaching mechanisms in hydrated cement pastes*, in Mechanisms of Chemical Degradation of Cement-Based Materials, E & FN Spon, pp. 38-49.

[10] Bentz, D.P. (1997), *Three-dimensional computer simulation of cement hydration and microstructure development*, Journal of the American Ceramic Society, Vol. 80, No. 1, pp. 3-21.

[11] Bentz, D.P. (2000), *CEMHYD3D: A three-dimensional cement hydration and microstructure development modelling package. Version 2.0*, NISTIR 6485, U.S. Department of Commerce.

[12] Marchand, J. (2000), *Modeling the behavior of unsaturated cement systemsexposed to aggressive chemical environments*, Materials and Structures, (in press).

[13] Samson, E. (2000), *Modeling ion transport mechanisms in unsaturated cement systems*, Ph. D. thesis, Department of Civil Engineering, Laval University, Canada, (in preparation).

[14] Bentz, D.P., Garboczi, E.J. (1992), *Modelling the leaching of calcium hydroxide from cement paste: Effects on pore space percolation and diffusivity*, Materials and Structures, Vol. 25, pp. 523-533.

[15] Bentz, D.P., Stutzman, P.E. (1994), *SEM analysis and computer modelling of hydration of portland cement particles*, in Petrography of Cementitious Systems, ASTM STP-1215, Ed. S.M. DeHayes and D. Stark, American Society for Testing and Materials, Philadelphia, pp. 60-73.

[16] Garboczi, E.J. (1998), *Finite element and finite difference programs for computing the linear electric and elastic properties of digitial images of random materials*, NISTIR 6269, U.S. Department of Commerce, (see http://ciks.cbt.nist.gov/monograph/, Chapter 2).

[17] Garboczi, E.J., Bentz, D.P. (1992), *Computer simulation of the diffusivity of cement-based materials*, Journal of Materials Science, Vol. 27, pp. 2083-2092.

[18] Bentz, D.P., Jensen, O.M., Glasser, F.P., Coats, A.M. (2000), *Influence of silica fume on diffusivity in cement-based materials - Part I: Experimental and computer modelling studies on cement pastes*, Cement and Concrete Research, Vol. 30, pp 953-962.

[19] Bear, J., Bachmat, Y. (1991), *Introduction to modeling of transport phenomena in porous media*, Kluwer Academic Publishers, The Netherlands.

[20] Samson, E., Marchand, J., Beaudoin, J.J. (1999), *Describing ion diffusion mechanisms in cement-based materials using the homogenization technique*, Cement and Concrete Research, Vol. 29, No. 8, pp.1341-1345.

[21] Helfferich, F. (1962), *Ion exchange*, McGraw-Hill, New York, USA, 624 p.

[22] Samson, E., Lemaire, G., Marchand, J., Beaudoin, J.J., (1999), *Modeling chemical activity effects in strong ionic solutions*, Computational Materials Science, Vol. 15, pp. 285-294.

[23] Samson, E., Marchand, J., Robert, J.L., Bournazel, J.P. (1999), *Modeling the mechanisms of ion diffusion transport in porous media*, International Journal of Numerical Methods in Engineering, Vol. 46, pp. 2043-2060.

[24] Pel, L. (1995), *Moisture transport in porous building materials*, Ph. D. thesis, Eindhoven University of Technology, The Netherlands, 125 p.

[25] Taylor, H.F.W. (1990), *Cement chemistry*, Academic Press Inc., San Diego, USA.

[26] Longuet, P., Burglen, L., Zelwer, A. (1980), *The liquid phase of hydrated cement*, Publication Technique CERILH, Vol. 219, (in French).

[27] Jacobsen, S., Marchand, J., Boisvert, L. (1996), *Effect of cracking and healing on chloride transport in OPC concrete*, Cement Concrete Research, Vol. 26, No. 6, pp. 869-882.

[28] Diamond, S. (1981), *Effects of two Danish fly ashes on alkali contents of pore solutions of cement fly ash pastes*, Cement and Concrete Research, Vol. 11, No. 2, pp. 383-390.

[29] Hazrati, K. (1995), *Investigation of the mechanisms of moisture transport by capillary suction in ordinary and high-performance cement-based materials*, Ph. D. thesis, Laval University, Canada, 205 p.

[30] Revertegat, E., Richet, C., Gegout, P. (1992), *Effect of pH on the durability of cement pastes*, Cement and Concrete Research, Vol. 22, pp. 259-272.

[31] Kuntz, M., Mareschal, J.C., Lavallee, P. (1997), *Numerical estimation of the effective conductivity of heterogeneous media with a 2D cellular automaton fluid*, Geophysical Research Letters, Vol. 24, No. 22, pp. 2865-2868, online at: http://www.agu.org/GRL/articles/97GL52856/ GL136W01.html.

Presentation: Sidney Diamond

Vagn Johansen and Niels Thaulow

CALCIUM HYDROXIDE IN CEMENT MATRICES:
PHYSICO-MECHANICAL AND PHYSICO-CHEMICAL CONTRIBUTIONS

J.J. Beaudoin
Institute for Research in Construction
National Research Council
Montreal Road, Bldg. M-20
Ottawa, Ontario K1A 0R6 Canada

ABSTRACT
Calcium hydroxide (CH) represents a significant volume of the products formed from the cement-water reaction. The extent to which the CH phase contributes to the engineering integrity and volume stability of cement-based binders is however moot. Evidence in support of the view that calcium hydroxide has a significant role in determining the mechanical performance and volume stability of cementitious materials is presented.

The elastic properties and fracture behavior of CH and CH – C-S-H mixtures are described. Volume stability of CH is described with reference to water sorption isotherms (CH) and length change measurements in various solvents and salt solutions.

A dissolution-based expansion mechanism in which CH plays a major role is described for the behavior of cement binders in aggressive media.

INTRODUCTION
Hydrated Portland cement paste contains two major constituents – calcium silicate hydrate and calcium hydroxide. The latter can be present in amounts up to 26% by volume. The principal binding phase, calcium silicate hydrate (C-S-H) has been extensively studied. The properties and behavior of calcium hydroxide (either as a pure phase or in mixtures of cementitious solids) have been investigated to a much lesser extent. The importance of the role of calcium hydroxide (CH) in the development of the mechanical, physical and chemical behaviors of cementitious systems has often been understated and can be considered moot[1]. This has been a result of

difficulties associated with the separation of C-S-H and CH (in hydrated paste) without affecting the integrity of the composite system.

The role of CH (in typical cement-based materials) with respect to descriptors of mechanical performance and durability is treated in this paper. The experimental evidence (obtained at the National Research Council, Canada) describing the physico-mechanical and physico-chemical contributions of CH in cement binders is collated and discussed.

The contribution of CH is elucidated using the following approach. Reference to sorption phenomena (with water as adsorbate) and extension mechanisms are made to validate the use of compacted specimens as representative solid bodies.

Compacted systems comprised of CH and CH/C-S-H mixtures are used as model systems to determine the potential contribution of CH to mechanical behavior and volume change stability in various chemical environments. This is a precursor to discussion of solvent exchange phenomena of CH compacts in methanol and isopropanol. This is followed by the presentation of length change (volume instability) behavior of CH in aggressive salt solutions.

The microstructure and properties of paste systems including undisturbed CH-depleted pastes are also described to provide evidence of a significant role for CH in the systems investigated.

SORPTION PHENOMENA

Sorption isotherms (eg. length change) in addition to providing surface-chemical data contain information pertaining to the elastic response of the sample[2]. The isotherm for CH compacts exhibits large primary and secondary hysteresis (Figure 1). Length change on second adsorption is reduced and the desorption curve is co-incident with the adsorption curve. There is large irreversible shrinkage on desorption. The length change – mass change curve (not shown) is linear up to $p/p_o = 0.35$. This is similar to that observed for hydrated portland cement paste. The length change on wetting is often referred to as the Bangham effect. It is due to the reduction of the solid surface energy resulting from the physical interaction of the surfaces with water molecules. The length change is directly proportional to the free energy change.

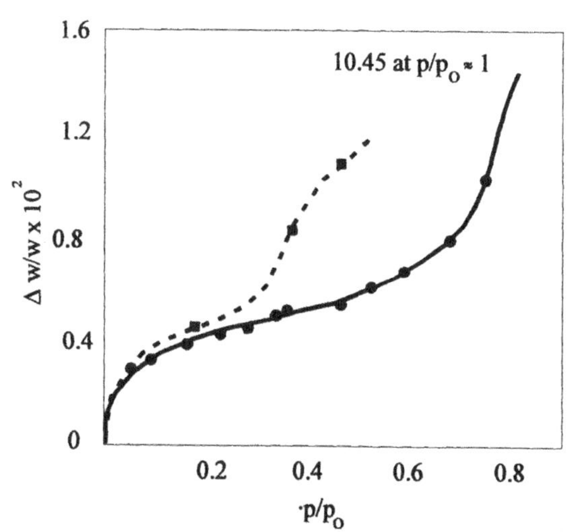

Figure 1 - Sorption isotherm of water on $Ca(OH)_2$ (adapted from reference2).

• — • adsorption ■ — ■ desorption

The change in free energy on adsorption (ΔF) can be determined from the Gibbs adsorption relation. A linear relation between length change and ΔF exists up to $p/p_o = 0.40$. The proportionality constant (λ) is related to the modulus of elasticity of the material through the relation $E = \rho\sigma/\lambda$ where E is the elastic modulus of the solid, ρ is the density and σ the surface area. The value of E calculated from sorption experiments is about 3.8 GPa. E values obtained from mechanical measurement of porous compacts (33% porosity) range from 3.4 GPa to 4.7 GPa. The values are of similar magnitude. Hydrated portland cement compacts gave values of 5.9 to 19.0 GPa depending on porosity values.

It is argued that adsorption – length change characteristics of CH at relative pressures of water vapor up to 0.40 are similar to those of other inert adsorbents. The Gibbs adsorption and Bangham equations can be invoked to explain behavior. Irreversible shrinkage that occurs during sorption – desorption cycling can be explained by dissolution at points of contact and ionic diffusion away from these sites. On desorption (especially at low humidities) crystallites are pulled together into the holes (created by dissolution processes) by van der Waal's forces. The menisci force may have a role at higher relative vapor pressures. Secondary hysteresis observed in the isotherms is attributed to trapping of water in pockets created by the action of van der Waal's forces.

CHEMICAL INTERACTIONS

Carbonation

Shrinkage of CH (lime) compacts results from exposure to carbon dioxide[3]. Experiments on compacts of bottle-hydrated cement show that carbonation of the 'combined-lime' is at least as rapid and extensive as the carbonation of the 'free lime' (Figure 2). Swenson and Sereda postulated that carbonation of the free lime occurs via a through solution process at points of contact where the material is most strained and solubility is the greatest. The reaction results in a local buildup of water. The carbonated product coats the surfaces and retards the reaction allowing moisture to diffuse away. Cycles of induced drying occur as a result of this transport causing the coating to crack and the cycle to repeat. Carbonation shrinkage of the C-S-H is considered to be related to the dehydration and polymerisation of the hydrous silica product of carbonation.

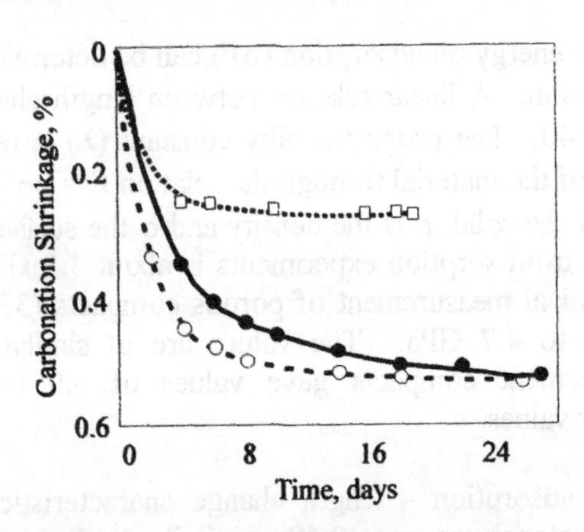

Figure 2 - Carbonation shrinkages of cement paste and compacts of hydrated cement and lime. Approximately equal porosity 100% CO_2 at 50% RH (adapted from reference 3). ○ Bottle-hydrated cement compacted at 70,000 lb • Cement paste, water/cement ratio 0.45 ◘ Calcium hydroxide compacted at 50,000 lb.

The carbonation process is a classic example of the contribution of both the CH and C-S-H phase to the volume stability of practical binders.

Solvent Exchange

The solvent exchange process has been considered as an alternative drying procedure for porous cement systems as most conventional drying procedures significantly affect the microstructure. Solvent exchange methods are believed to 'preserve' the microstructure to some extent. It has been shown by the author and coworkers that CH compacts exhibit similar tendencies to C-S-H when exposed to methanol and isopropanol[4].

A brief discussion of the length change of CH compacts exposed to water, methanol and isopropanol serves to illustrate further the dual role of C-S-H and CH in hydrated cement systems. Compacts of CH immersed in water expand as previously indicated. Expansion is attributed to the Bangham effect (up to a mass change of 0.70% and length change of 0.10%) and solution/precipitation at points of contact involving diffusion to and recrystallization at other sites.

Most of the expansion of dry CH in isopropanol is probably due to the Bangham effect as a value of about 0.12% would be expected. Calcium hydroxide is insoluble in isopropanol and a dissolution/precipitation mechanism for length change is not considered tenable. Length change of partially saturated CH compacts immersed in isopropanol is significantly different. There is an immediate expansion followed by a rapid and progressive contraction (Figure 3). This may be due to some surface interaction between isopropanol and CH. This cannot be explained by the Bangham effect as the surface free energy balance would favor a reduced length change. The progressive contraction can be explained as a shrinkage resulting from a bulk removal of water by isopropanol where the rate of removal exceeds the rate of replacement.

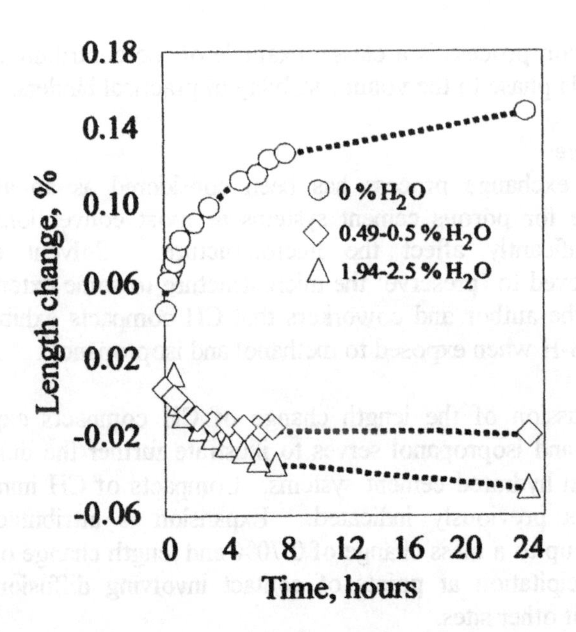

Figure 3 - Length change of $Ca(OH)_2$ compacts of varying water content immersed in isopropanol.

The CH compacts (dry or partially saturated) undergo expansion during solvent replacement with methanol (Figure 4). It suggested that this is due to a combination of the Bangham effect and some type of chemical interaction. It would appear (for the partially saturated CH compacts) that the expansive nature of the chemical interaction overrides any tendency for contraction due to solvent replacement action or changes in free energy due to the replacement of water with solvent molecules at the solid surface. The similarities in the basic trends of the length change – time curves for the methanol and isopropanol exchanged CH and cement paste systems (expansion and contraction) are striking and reinforce the view that CH in hydrated cements contributes in a major way to volume change stability.

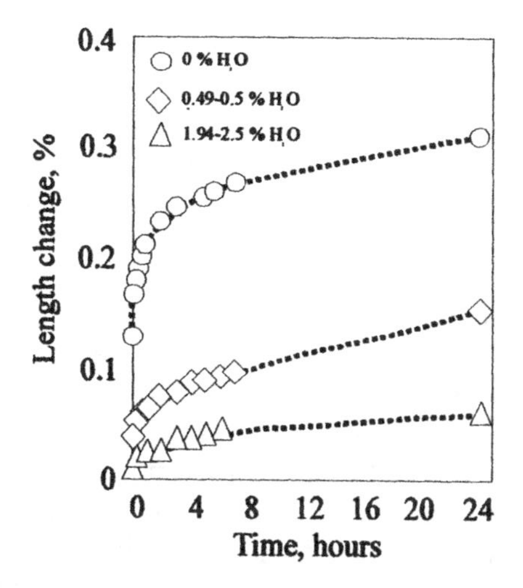

Figure 4 - Length change of $Ca(OH)_2$ compacts of varying water content immersed in methanol.

Volume Stability in Salt Solutions

The length change behavior of compacted CH specimens in various salt solutions including $MgCl_2$ and $MgSO_4$ suggests that the formation of new crystalline phases in cement systems may not be (in itself) an à priori condition for expansion[5]. Evidence for this is provided in the following brief discussion.

The CH compacts expanded in all the test solutions (including saturated lime solution) confirming that in the majority of cases the dissolution of this important component of hydrated cement systems is intrinsically expansive (Table I).

Table I. Expansion of CH Compacts in Various Solutions*

Solution	Expansion, %
$MgCl_2$	0.80
NaCl	0.48
$MgSO_4$	0.05
Distilled Water	0.52
Lime Water	0.30
NH_4NO_3	0.85

* Concentration of salt solutions (180 g/l); lime water is a saturated solution

The order of expansion was not in the order of respective solubilities. Immersion in the $MgSO_4$ solution gave the lowest expansion in spite of having the highest solubility with respect to calcium hydroxide dissolution. This can be explained by the possibility of competing forces between the dissolution of the calcium hydroxide (expansive) and the coating of calcium hydroxide particle surfaces with reaction products e.g. brucite and gypsum (contractive). This process is rate of dissolution and concentration dependent. It also likely depends on the nature of the precipitate. Hence the net length change effect varies with the solution type and results in larger observed expansions for immersion in $MgCl_2$ solution. It is clear that crystalline hydration products (in hydrated cement systems) such as calcium hydroxide are sources of potential expansion when immersed in chloride and sulfate solutions. The crystals of CH are in an initial state of strain which is released upon dissolution[6]. The results provide additional evidence of the importance of CH in assessing the durability of cement-based materials.

CH Depletion in Distilled Water and Partially Saturated Lime Solutions

Leaching of free $Ca(OH)_2$ (in distilled water) from saturated hydrated portland cement results in a slight expansion. Leaching interlayer and structural $Ca(OH)_2$ results in a slight contraction[7]. During drying of leached paste a large shrinkage occurs between 11% RH and final drying. This shrinkage increases significantly with the amount of $Ca(OH)_2$ removed by leaching. The large shrinkage after removal of 17% calcium hydroxide indicates that some CH was removed from between the sheets while free calcium hydroxide was being removed. This is consistent with the explanation that removal of Ca^{++} ions from between C-S-H sheets results in greater collapse of the sheet and a larger amount of water re-entry on rewetting. These simple experiments underlie the importance of CH in providing structural stability to cement binders.

The effect of leaching (in partially saturated lime solutions) and overleaching on the microstructure of C-S-H can be seen in figure 5[8]. The term overleaching refers to the removal of calcium hydroxide from the C-S-H itself subsequent to the removal of all the free calcium hydroxide from the cement paste. The specimens were subjected to an excess pressure of helium in a vacuum cell. The leached and unleached specimens show somewhat similar results, both demonstrating an abrupt decrease of helium flow (at a pressure of 0.20 MPa) at about 7% mass loss indicative of a collapse of structure. The curves for the overleached samples show that helium flow increases continuously as water is removed.

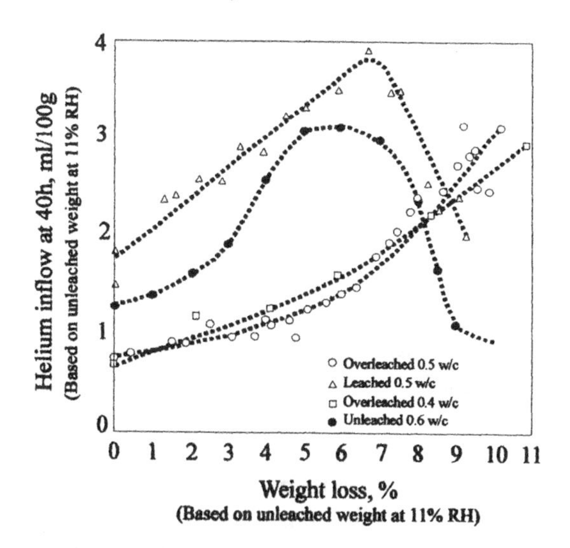

Figure 5 - Helium inflow at 40h as a function of weight loss for leached cement paste (adapted from reference 7).

PHYSICO-MECHANICAL BEHAVIOR

Fracture and Strength of C-S-H and CH Compacts

The C-S-H or C-S-H/CH mixtures were made in the form of thin plates (double-torsion geometry) for crack growth studies[9]. The porosity dependence of the critial stress intensity factor, K_c is illustrated in Figure 6. The data for all the C-S-H and C-S-H/CH specimens appear to be best described by a linear relationship independent of C/S ratio (0.68-1.49). The data for the pure CH is also described by a linear relationship.

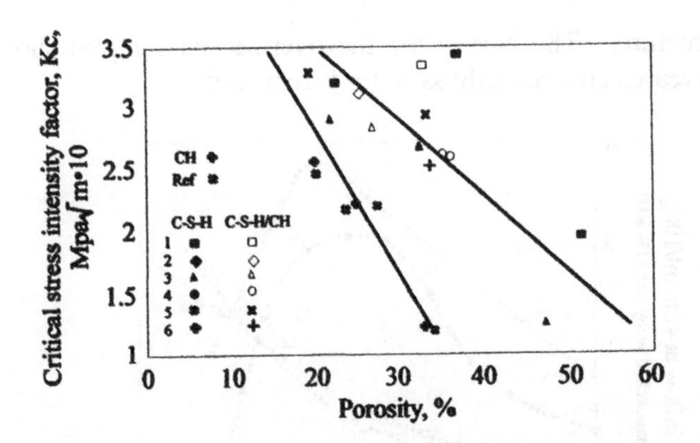

Figure 6 - Critical stress intensity factor, K_c vs porosity for compacted specimens of synthetic C-S-H, C-S-H/CH mixtures and CH (adapted from reference 9).

The K_c values for CH are lower than those for the C-S-H systems at equivalent porosity values. This is consistent with observations that K_c values of CH are lower than those of hydrated cement paste at the same porosity (Figure 7)[10]. Values of modulus of elasticity of portland cement paste and CH compacts are plotted in Figure 8. It is evident that values of modulus of elasticity for CH compacts are of the same order of magnitude as the values for portland cement paste. Flexural strength data for CH compacts range from 5.89 MPa (porosity = 34%) to 14.20 MPa (porosity = 21.0%). The values for CH compacts are of a magnitude similar to those for portland cement paste over a similar porosity range (i.e. 3.00 – 10.00 MPa; porosity = 17.0 – 30.0%).

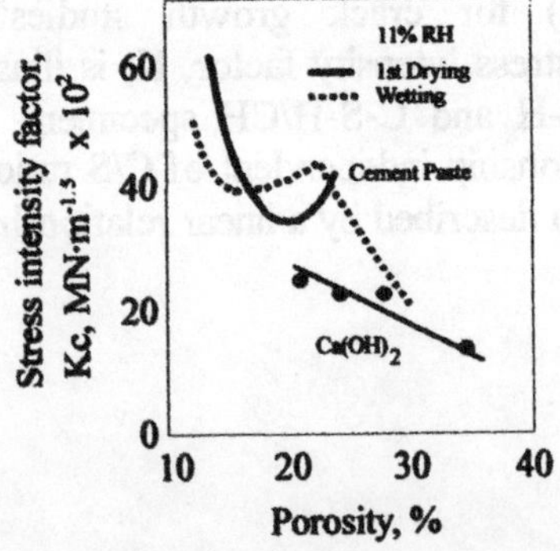

Figure 7 - Critical stress intensity factor vs porosity for $Ca(OH)_2$ compacts and Portland cement paste (adapted from reference 10).

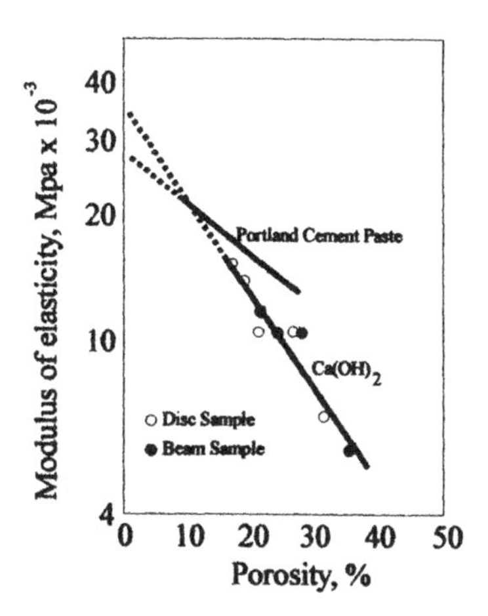

Figure 8 - Modulus of elasticity of $Ca(OH)_2$ compacts and portland cement paste vs porosity (adapted from reference 10).

CONCLUDING REMARKS

Similarities in the physico-chemical and physico-mechanical behavior between phase pure CH and hydrated cement paste (comprised largely of C-S-H and CH) suggest that both principal components play a significant role in determining engineering behavior. This is evident from estimates of modulus of elasticity of CH compacts based on both sorption phenomena and mechanical measurements. The similarities extend to volume change measurements associated with carbonation, solvent exchange and chemical phenomena in salt solutions. The latter provide an indication that the dissolution of CH crystals can be associated with expansive interactions in cement-based materials. This is confirmed through length change measurements during leaching experiments conducted on cement paste. The volume stability of the paste is affected by the leaching process (especially overleaching) as removal of lime from between the C-S-H sheets increases the potential for structural collapse and shrinkage.

The flexural strength and fracture parameter values of paste, C-S-H, CH and mixtures of C-S-H and CH are of a similar order of magnitude. This underlies the theme that consideration should be given to the mechanical and chemical stability of both principal phases in the design of durable concrete structures.

REFERENCES

[1]J. Marchand, J.J. Beaudoin and M. Pigeon, "Influence of Calcium Hydroxide Dissolution on the Engineering Properties of Cement-Based Materials" in Materials Science of Concrete, Edited by J. Marchand and J. Skalny, The American Ceramic Society, Westerville Ohio, 2000.

[2]V.S. Ramachandran and R.F. Feldman, "Length Change Characteristics of $Ca(OH)_2$ Compacts on Exposure to Water Vapour", Journal of Applied Chemistry, 17, 328-32 (1967).

[3]E.G. Swenson and P.J. Sereda, "Mechanism of the Carbonatation Shrinkage of Lime and Hydrated Cement"" Journal of Applied Chemistry, 18, 111-17 (1968).

[4]J.J. Beaudoin, P. Gu, J. Marchand, B. Tamtsia, R.E. Myers and Z. Liu, "Solvent Replacement Studies of Hydrated Portland Cement Systems: The Role of Calcium Hydroxide", Advanced Cement Based Materials, 8, 56-65 (1998).

[5]J.J. Beaudoin, S. Catinaud and J. Marchand, "Volume Stability of Calcium Hydroxide in Aggressive Solutions", Cement and Concrete Research (in press).

[6]J.E. Gillott and P.J. Sereda, "Strain in Crystals Detected by X-ray", Nature, 209 [5018] 34-36 (1966).

[7]R.F. Feldman and V.S. Ramachandran, "Length Changes in Calcium Hydroxide-Depleted Portland Cement Pastes", Il Cemento, 86 [2] 87-96 (1989).

[8]R.F. Feldman and V.S. Ramachandran, "Microstructure of Calcium Hydroxide-Depleted Portland Cement Paste 1: Density and Helium Flow Measurements", Cement and Concrete Research, 12 [2] 179-89 (1982.

[9]J.J. Beaudoin, P. Gu and R.E. Myers, "The Fracture of C-S-H and C-S-H/CH Mixtures", Cement and Concrete Research, 28 [3] 341-47 (1998).

[10]J.J. Beaudoin, "Comparison of Mechanical Properties of Compacted Calcium Hydroxide and Portland Cement Paste Systems", Cement and Concrete Research, 13 [3] 319-24 (1983).

THE STRENGTH AND FRACTURE OF CONCRETE: THE ROLE OF THE CALCIUM HYDROXIDE

Sidney Mindess
Department of Civil Engineering
University of British Columbia
Vancouver, B.C. V6T 1Z4, Canada

ABSTRACT

Calcium hydroxide crystals constitute about 20% to 25% of the volume of solids in hydrated cement paste. It is generally assumed that the calcium hydroxide is weaker than the C-S-H, thus providing a preferential fracture path. However, the role of the calcium hydroxide in determining the strength and fracture properties of concrete is in fact far from clear. In this literature review, the effects of the calcium hydroxide on both normal and high strength concretes are examined.

INTRODUCTION

In ordinary portland cement concretes, calcium hydroxide crystals (CH, or portlandite) make up about 20 to 25% of the volume of the hydrated cement paste (hcp). They grow wherever free space is available: largely within the capillary pore space, and in the interfacial transition zone (ITZ) at the cement-aggregate interface. Depending on the conditions under which the hydration takes place, the CH crystals may vary considerably in morphology, as described in detail by Bache *et al.* (1966), amongst others:

- Separate or aggregate crystals and crystallites scattered in the cement paste (Figs. 1-3);
- Layers covering the surfaces of aggregate particles, ranging in thickness from a few μm to more than 50 μm;
- In what were originally bleed water pockets on the undersides of coarse aggregate particles;

- In narrow spaces between aggregate particles;
- As CH shells of varying thickness precipitated *around* air-filled voids;
- and as large hexagonal crystals *within* air bubbles.

The size of the crystals ranges from submicroscopic to perhaps 200 μm.

It is interesting that, as long ago as 1887, Le Chatelier, in his Doctoral thesis, observed that CH (which he referred to as calcium hydrate) "crystallizes by surrounding and joining all the foreign substances which it contains." That is, it may also engulf other materials (*cf.* Fig. 5, below). He also noted the "relatively considerable dimensions of these crystals".

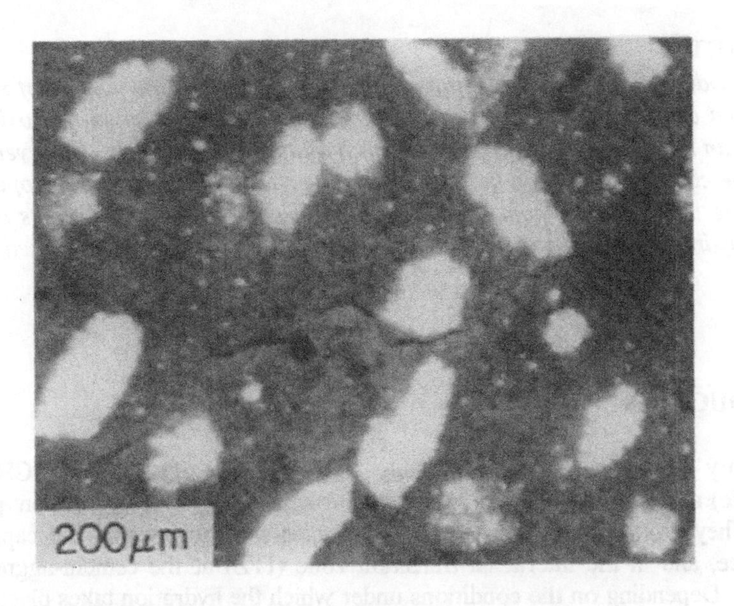

Fig. 1. Optical micrograph of CH in 1-day old hydrated cement paste – isolated crystals (Micrograph courtesy of J.F. Young).

MECHANICAL PROPERTIES OF CALCIUM HYDROXIDE

Generally, it is assumed that because CH has a much lower specific surface area than does C-S-H, its contribution to strength (through van der Waals forces) must also be low. That is, it is considered that CH is "weaker" than C-S-H. Unfortunately, however, there are virtually no data available on the mechanical properties of CH crystals, because it has not been possible to prepare single crystals of CH that are large enough to be tested mechanically. Yet, if we wish to understand the factors determining concrete strength, or to use mathematical models to predict the mechanical properties of cementituous systems, the mechanical properties of CH must somehow be obtained or estimated.

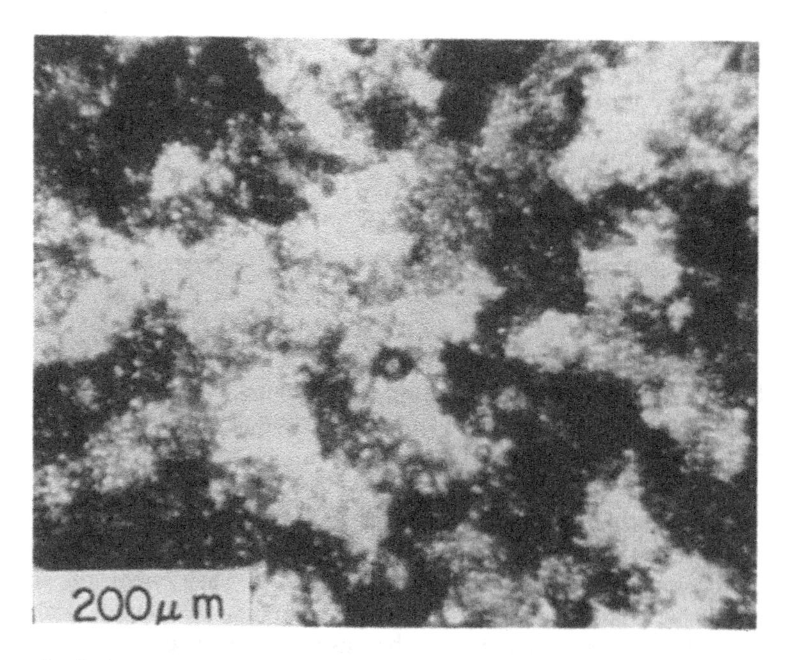

Fig. 2. Optical micrograph of CH in 21-day old hydrated cement paste, showing the spread of Young).the CH throughout the paste (Micrograph courtesy of J.F.

Such data as do exist have mostly been obtained on specimens prepared by compacting CH powder samples to different porosities, and then extrapolating back to "zero" porosity. The values reported in the literature for modulus of elasticity are given in Table 1. Not surprisingly, these data are somewhat inconsistent, since the lowest porosities obtained by compaction have been at least 18% or more. Nevertheless, it may be seen that, when extrapolated back to "zero porosity", the available data for CH are in the same range as those for hydrated cement or C_3S pastes. Only the "solid phase" (zero gel porosity) data of Helmuth and Turk (1966) give significantly higher E values.

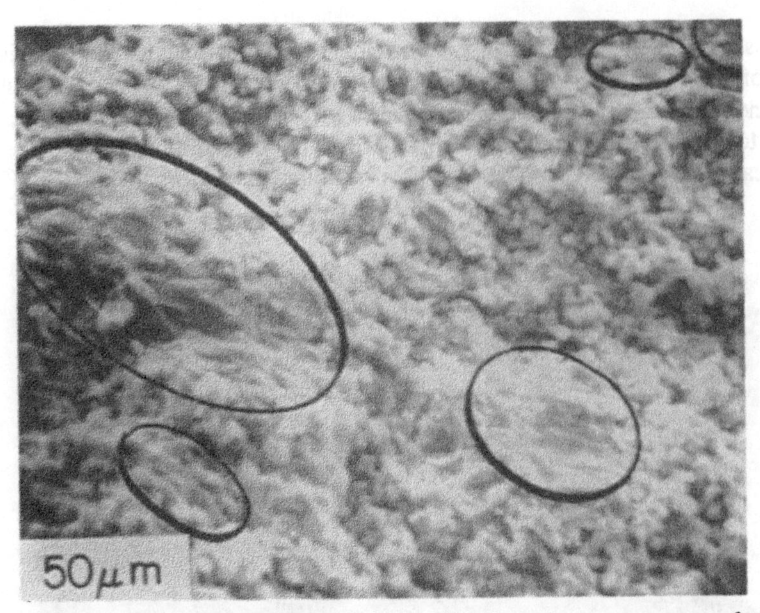

Fig. 3. SEM micrograph of CH in 3-day old hydrated cement paste; the circles enclose CH crystals, showing the characteristic striated appearance (Micrograph courtesy of J.F. Young).

The only available direct measurements of strength on CH crystals are the direct tensile tests of Baldie (1985). These tests were carried out on CH single crystals prepared from three different solutions. They were of columnar hexagonal shape, with both height and width in the range of 1mm to 2mm. There is considerable scatter in the results because of the experimental difficulties involved in testing very small crystals, and because of pre-existing flaws in the crystals. As well, the specimens were not all loaded in the same orientation with respect to the basal plane. The results do, however, provide at least a rough "lower limit" of the strength of CH crystals. For the 11 specimens tested, the tensile strength values ranged from 0.49 MPa to 2.19 MPa with an average value of 1.13 MPa. This is somewhat lower than the typical tensile strengths of hcp, which generally are in the range of about 2MPa to perhaps 10MPa, but, as stated before, Baldie's results are really represent a lower bound to the strength range in CH.

The only other readily available strength data for CH (obtained on compacted powders) are those reported by Beaudoin (1983). He found that, for the porosity range examined (21-34%), the *flexural strengths* of the CH compacts were in the same range as those of the hydrated cement paste. The microhardness of CH, (which may empirically be related to strength) was found to be higher than that of hcp.

Thus, on the basis of the available elastic modulus and strength data, there is no reason to believe that the CH is particularly weaker than the hcp in ordinary concretes. Indeed, it has long been known that the leaching out of the CH from concrete leads to considerable strength loss (e.g. Terzaghi, 1944). Even earlier, Tremper (1931) had shown that when lime is leached out of concrete, the loss of concrete strength (as a percentage of the original strength) was approximately equal to two times the loss of lime (as a percentage of the original lime content). More recently, Carde *et al.* (1996, 1997) found that when concrete was leached by ammonium nitrate, "the dissolution of this calcium hydroxide is the essential parameter governing both decrease in strength and increase in porosity."

FRACTURE PROPERTIES OF CALCIUM HYDROXIDE

In pure C_3S pastes, it has been shown by Berger (1972) that, at early ages, cracks propagate preferentially *around* CH crystals (Fig. 4). By the age of three days, some CH crystals even seem to act as crack arrestors (Fig.5) However, at later ages, beyond about 1 month, the cracks will propagate more directly *through* both the CH crystals and the C-S-H.

In concretes, it is generally found (e.g. Dhir *et al.*, 1996; Bentz *et al.*, 1995) that the interfacial region between the cement paste and aggregate particles generally contains massive oriented crystals of CH. Dhir et al (1996) go on to say that "the nature of these crystals influences the strength of the cement paste-aggregate bond, which in turn affects the strength of the concrete as a whole." Bonen (1994) has shown that the CH forms irregular coatings around aggregate particles, ranging in thickness from about 2-3 μm to perhaps 8 μm. Similar values have been found by Dhir *et al.* (1996) and by Mouret *et al.* (1999), though as stated earlier in Bache *et al.* (1966) found CH thicknesses of up to 50 μm. According to Mouret *et al.* (1999) the individual CH crystals are 10-15 μm in size, and may have different orientations with respect to the physical interface, with their c-axes either parallel or perpendicular (basal planes perpendicular or parallel, respectively) to the interface.

Table I. Estimations of the Elastic Modulus of CH

Reference	E (GPa)	Basis of Estimate
CH		
Ramachandran & Feldman (1967)	3.5 - 4.5	CH powder compacted to 136 MPa.
Swenson & Sereda (1968)	8.84	CH powder compacted to 340 MPa.
Beaudoin (1983)	35.24	CH powders compacted at various pressures, extrapolated back to "zero porosity".
Wittmann (1986)	48	Compacted CH powders, extrapolated back to "zero porosity".
Monteiro & Chang (1995)	39.77 - 44.20	Calculated from values of the bulk modulus and shear modulus.
Cement Pastes (for comparison)		
Beaudoin (1983)	28	Both compacted and *in situ* hydrated cement pastes, "zero porosity".
Helmuth & Turk (1966)		
Hydrated C_3S Paste	46.9 gel phase 95 solid phase	*In situ* hydrated C_3S, extrapolated back to "zero porosity".
Hydrated cement paste	31.0 - 31.7 gel phase 74.5 - 80.0 solid phase	*In situ* hydrated cement, extrapolated back to "zero porosity".

Fig. 4 Crack growth around CH crystals in 14-day old hydrated C_3S paste. The grey areas are the CH crystals. The cracks appear to be largely in the C-S-H phase (Photograph courtesy of R. L. Berger).

It is these **oriented CH crystals at the interface** that are primarily responsible for the view that CH "weakens" the concrete, because microcracks can grow more easily along the cleavage planes of the CH (Marchese, 1977; Walsh *et al.*, 1974). Indeed, while the mechanical properties of CH appear to be similar to those of hcp, Beaudoin (1983) also found that the **fracture parameters** of CH are significantly lower than those of mature hcp, in terms of fracture toughness and fracture energy. Many studies have shown large CH crystals exposed at the interfacial fracture surfaces, presumably as a result of this type of cleavage. Large CH crystals are, of course, also found in the large capillary air voids throughout the bulk paste. However, because these CH crystals are not oriented in any particular direction and are dispersed throughout the concrete, they do not constitute a preferential fracture plane.

As well, when smaller CH crystals are incorporated within the irregular C-S-H structure (Fig. 6), they do not constitute points of crack initiation, since they too are randomly oriented, and are in any event as strong as the surrounding paste.

MITIGATION OF THE EFFECTS CH

As we have seen above, CH does not particularly affect the *strength* of concrete, *per se*. However, when it occurs as massive crystals at the cement-aggregate interface, fracture will take place preferentially along the cleavage planes. How can this weakness be mitigated?

There are basically two approaches which may be taken. Most commonly, particularly in the production of high strength concrete, silica fume is used at an addition rate of about 10-15% by weight of the cement. The pozzolanic reaction between the silica fume and the CH helps to consume at least some of the CH, transforming it to C-S-H; as well, the much lower (cementitious material) ratio in high strength concrete leaves much less space at the interface within which CH crystals can form. This leads to a much denser and stronger interface, with no obvious planes of weakness, as has been shown by Bentur and Cohen (1987), Bentur (1991) and Odler and Zurz (1988), amongst others. Of course, the silica fume does other things at the interface as well, eliminating many of the larger pores, and making the structure more homogeneous. It may also modify the rheological properties of the fresh concrete in such a way that bleeding is reduced at the interface. All of these effects contribute to the production of high strength concrete, though there is no consensus as to their relative importance (Aitcin and Mindess, 1998).

The other approach would be to use chemical admixtures to alter the morphology of the CH that does form, transforming it to smaller, more compact and more evenly distributed crystals. This too may lead to a strengthening of the interfacial zone. Most likely, some combination of these two approaches (i.e., silica fume plus chemical admixtures) will be found to be most effective.

CONCLUSIONS

Calcium hydroxide appears to be about as strong as the C-S-H. However, because of its morphology at the cement-aggregate interface, it provides a plane of weakness along which cracks can propagate more easily, this providing the "weak link" in ordinary concrete. Consuming this CH by the pozzolanic reaction with silica fume, or altering its morphology by the use of chemical admixtures, can lead to considerable strengthening of the cement-aggregate bond, and hence of the concrete itself.

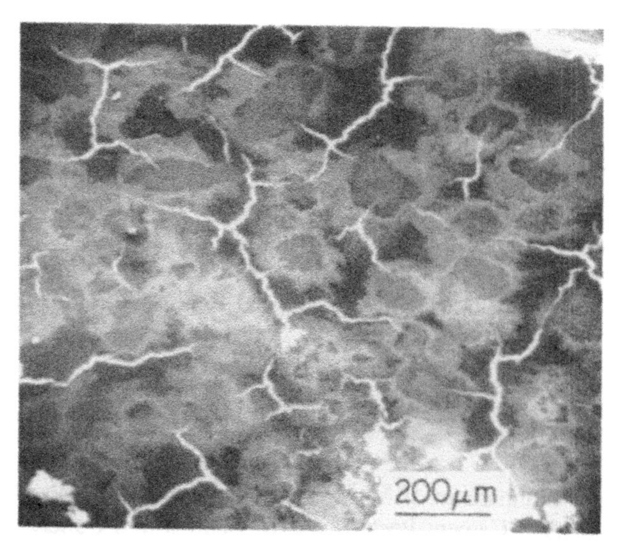

Fig. 5 - Cracks "arrested" by CH crystals (dark gray) in 14-day old hydrated C_3S paste. The light gray areas represent CH+ engulfed C-S-H; the cracks appear to run largely in the C-S-H phase (micrograph courtesy of R.L Berger).

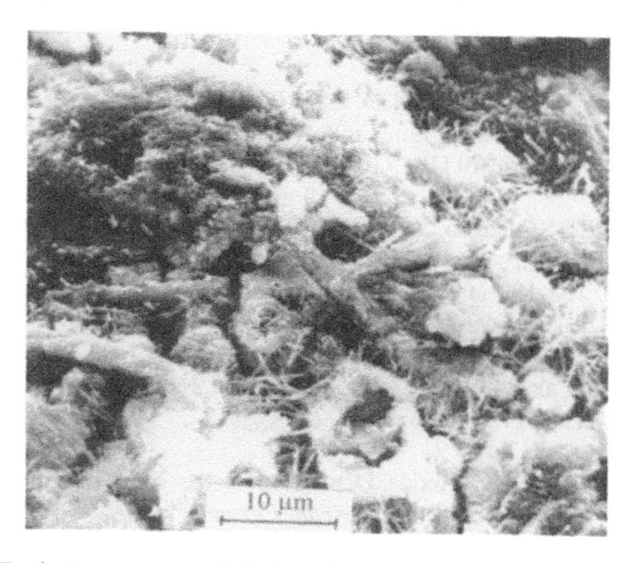

Fig. 6 – Typical structure of hydrated cement paste (SEM courtesy of S. Diamond)

REFERENCES

P.C. Aitcin and S. Mindess, "High-Performance Concrete: Science and Applications," pp. 477-511 in J. Skany and S. Mindess (eds.), *Materials Science of Concrete V*, The American Ceramic Society, Westerville, Ohio, 1998.

H.H. Bache, G.M Idorn, P.Nepper-Christensen and J. Nielsen, "Morphology of Calcium Hydroxide in Cement Paste," pp. 154-174 in *Symposium on Structure of Portland Cement Paste and Concrete*, Special Report 90, Highway Research Board, Washington, D.C., 1966.

K.D. Baldie, Ph.D. Dissertation, University of London (Imperial College), 1985.

J.J. Beaudoin, "Comparison of Mechanical Properties of Compacted Calcium Hydroxide and Portland Cement Paste System," *Cement and Concrete Research*, **13** [3] 319-324 (1983).

A. Bentur, "Microstructure, Interfacial Effects, and Micromechanics of Cementitious Composites; pp. 523-549 in S. Mindess (ed.), *Advances in Cementitious Materials*, Concrete Transactions **16**, The American Ceramic Society, Westerville, Ohio, 1991.

A. Bentur and M.D. Cohen, "The Effect of Condensed Silica Fume on the Microstructure of the Interfacial Zone in Portland Cement Mortar," *Journal of the American Ceramic Society*, **70** [10] 738-742 (1987).

D.P. Bentz, E. Schlangen and E. J. Garboczi, "Computer Simulation of Interfacial Zone Microstructure and Its Effect on the Properties of Cement-Based Composites," pp. 155-199 in J. Skalny and S. Mindess (eds.), *Materials Scietnce of Concrete IV*, The American Ceramic Society, Westerville, Ohio 1995.

R.L. Berger, "Calcium Hydroxide: Its Role in the Fracture of Tricalcium Silicate Paste," *Science* **175**, 626-629 (1972).

D. Bonen, "Calcium Hydroxide Deposition in the Near Interfacial Zone in Plain Concrete," *Journal of the American Ceramic Society*, **77** [1] 193-196 (1994).

C. Carde and R. Francois, "Effect of the Leaching of Calcium Hydroxide from Cement Paste on Mechanical and Physical Properties," *Cement and Concrete Research*, **27** [4] 539-550 (1997).

C. Carde, and R. Francois and J.M. Torrenti, "Leaching of Both Calcium Hydroxide and C-S-H from Cement Paste: Modeling the Mechanical Behavior," *Cement and Concrete Research*, **26** [8] 1257-1268 (1996).

R.K. Dhir, P.C. Hewlett and T.D. Dyer, "Influence of Microstructure on the Physical Properties of Self-Curing Concrete," *ACI Materials Journal*, **93** [5] 465-471 (1996).

R.A. Helmuth and D.H. Turk, "Elasic Moduli of Hardened Portland Cement and Tricalcium Silicate Pastes: Effect of Porosity:" pp. 135-144 in *Symposium on Structure of Portland Cement Paste and Concrete,* Special Report 90, Highway Research Board, Washington, D.C., 1966.

H. Le Chatelier, "Experimental Researches on the Constitution of Hydraulic Mortars" (1887); translated by J.L. Mack; McGraw Publishing Company, New York, 1905.

B. Marchese, "SEM Topography of Twin Fracture Surfaces of Alite Pastes 3 Years Old,"*Cement and Concrete Research*, **7** [1] 9-18 (1977).

P.J.M. Monteiro and C.T. Chang, "The Elastic Moduli of Calcium Hydroxide," *Cement and Concrete Research*, **25** [8] 1605-1609 (1995).

M. Mouret, A. Bascoul and G. Escadeillas, "Microstructural Features of Concrete in Relation ot Initial Temperature – SEM and ESEM Characterization," *Cement and Concrete Research*, **29** [3] 369-375 (1999).

I. Odler and A. Zurz, "Structure and Bond Strength of Cement – Aggregate Interfaces," *Advances in Cement Research*, **1** [4] 21-27 (1988).

V.S. Ramachandran and R.F. Feldman, "Length Change Characteristics of Calcium Hydroxide Compacts on Exposure to Water Vapor," *Journal of Applied Chemistry*, **17** [11] 328-332 (1967).

E.G. Swenson and P.J. Sereda, "Mechanism of the Carbonation Shrinkage of Lime and Hydrated Cement," *Journal of Applied Chemistry*, **18** [4] 111-117 (1968).

R.D. Terzaghi, "Concrete Deterioration in a Shipway," *Journal of the American Concrete Institute,* **44** [6] 977-1008 (1944).

B. Tremper, "The Effect of Acid Waters on Concrete," *Journal of the American Concrete Institute*, **28** [9] 1-32 (1931).

D. Walsh, M.A. Otooni, M.E. Taylor, Jun. and M.J. Marcinkowski, "Study of Portland Cement Fracture Surfaces by Scanning Electron Microscopy Techniques," *Journal of Materials Science*, **9**, 423-429 (1974)

F.H. Wittmann, "Estimation of the Modulus of Elasticity of Calcium Hydroxide," *Cement and Concrete Research*, **16** [6] 971-972 (1986).

EFFECT OF CALCIUM HYDROXIDE ON THE PERMEABILITY OF CONCRETE

Nataliya Hearn
University of Windsor, Windsor, Ontario, Canada

ABSTRACT

The healing of concrete encompasses different mechanisms, depending on the environment of exposure. Healing due to carbonation, marine exposure or blocking of the pore structure is all connected to the presence and reaction with calcium hydroxide. Field experience in cracked water-retaining structures has shown decrease in flow through cracks. Similar behaviour is observed in micro cracked concrete, where permeation characteristics of concrete have been known to decrease with the progress of the permeability test. Several possible mechanisms were investigated, including the presence of calcium hydroxide. It was found that the self-sealing property is related to the presence and dissolution of the calcium hydroxide from the pore structure into the pore system.

INTRODUCTION

Many simultaneous mechanisms contribute to the evolution of the cementitious matrix; among them hydration, swelling, dissolution, and re-deposition of solute, etc. This study focused on evaluating effects of dissolution of calcium hydroxide on the permeation properties of the cementitious matrix. These changes in the matrix were evaluated through the changes in the water permeability measurements. The concrete tested was cured in water for 26 years, so that terminal hydration level was reached, and thus hydration was not significant in the permeability analysis. The samples were tested under saturated conditions, so that swelling of the microstructure had already taken place [1]. The following discussion addresses the extent to which permeability can be affected by the dissolution of calcium hydroxide.

BACKGROUND

The fundamental permeation characteristics of concrete are a function of the continuity of the pore structure, which has been frequently defined by the largest continuous pore radius [2]. In concrete, the largest continuous pore radius is initially defined by the w/c ratio and the type and amount of supplementary cementing materials. After setting, the largest continuous pore radius becomes progressively reduced as the hydration products fill in the available pore space. In well-hydrated systems, even at w/c ratios of 0.55, the pore structure can easily become discontinuous, so that permeability drops to immeasurable amounts under standard testing procedures, as has been demonstrated by modeling work of Garboczi and Bentz [2]. With exposure, the process of continuing hydration can be counteracted by development of cracking. The cracking can be on a macro-scale, creating discontinuities in the concrete elements, or on a micro-scale, creating continuous channels in the concrete's microstructure.

In the presence of water, concrete has the ability to seal developed cracks. This internal resistance is provided by the presence of calcium hydroxide in the cementitious structure. The most obvious examples of this sealing are the "autogenous" sealing of cracks in water-retaining structures, culverts and pipes [3,4,5]. The white crystalline precipitate which fills and forms scar tissue over cracks, is caused by either of the two processes:

- reaction between calcium hydrogen carbonate ($Ca(HCO_3)_2$) from water and calcium hydroxide ($Ca(OH)_2$) from concrete, or
- carbonation of calcium hydroxide dissolved from concrete upon exposure to the atmosphere.

In marine environment the sealing mechanism is somewhat different as it is mostly controlled by the magnesium salts present in seawater and affects only the outer layer of concrete. The reactions involved are:

- calcium hydroxide from concrete and Magnesium salts the seawater forming a layer of relatively insoluble brucite, and
- calcium carbonate and magnesium salts resulting in deposition of aragonite[6]

In regular concrete the internal micro cracks will also undergo self-sealing process, mostly due to the presence of calcium hydroxide in the pore structure. This presentation is an attempt to describe the experimental results evaluating the effect of $Ca(OH)_2$, on permeability of concrete.

TESTING AND RESULTS

Tested Samples

Concrete samples were cast into 100Øx 200mm cylinders in 1964, and stored in water since casting. The concrete mix proportions are given in Table 1.

Table 1. Mix Proportions of Concrete Samples

W/C	0.9
A/C	7.5
Constituents (ratio by wt.)	
RHPC	1
Fines	0.3
Builders' sand	2.02
Builders' gravel	5.17
Water	0.9
Date cast	8/1/1964
Age at testing	25-26 years

In order to assess the effect of calcium hydroxide on the permeability of concrete samples, the following tests were conducted: water and propan-2-ol permeability, pore solution analysis, and SEM examination of the fracture surfaces.

Water Permeability

Concrete cores were removed from the water bath in which they were stored, and cut into 50mm thick disks for water permeability testing. The water permeability tests were conducted on these samples after cutting. After testing, these samples were oven dried at 105°C until constant weight was reached. After oven drying, concrete samples were re-saturated and re-tested for water permeability.

Water permeability apparatus is shown in Figure 1 and has been described in previous papers [7]. Testing was conducted using 0.7 MPa pressure applied at the top face of the test specimen. The bottom face, at which the outflow was measured, was kept at atmospheric pressure. Permeability testing was thus conducted using Darcian flow parameters, with the intrinsic permeability calculated using the following relationship:

$$D = \frac{\mu}{g\rho}\,\frac{Q}{A}\,\frac{\partial p}{\partial x}$$

where: D = intrinsic permeability coefficient (m^2)
ρ = density of the permeant (kg/m^3)
g = gravitational acceleration (9.81 m/s^2)
μ = viscosity (Pa.s)
Q = rate of flow (m^3/s)
A = cross-sectional area of the sample (m^2)
∂p = pressure drop in the x (m) direction (Pa)

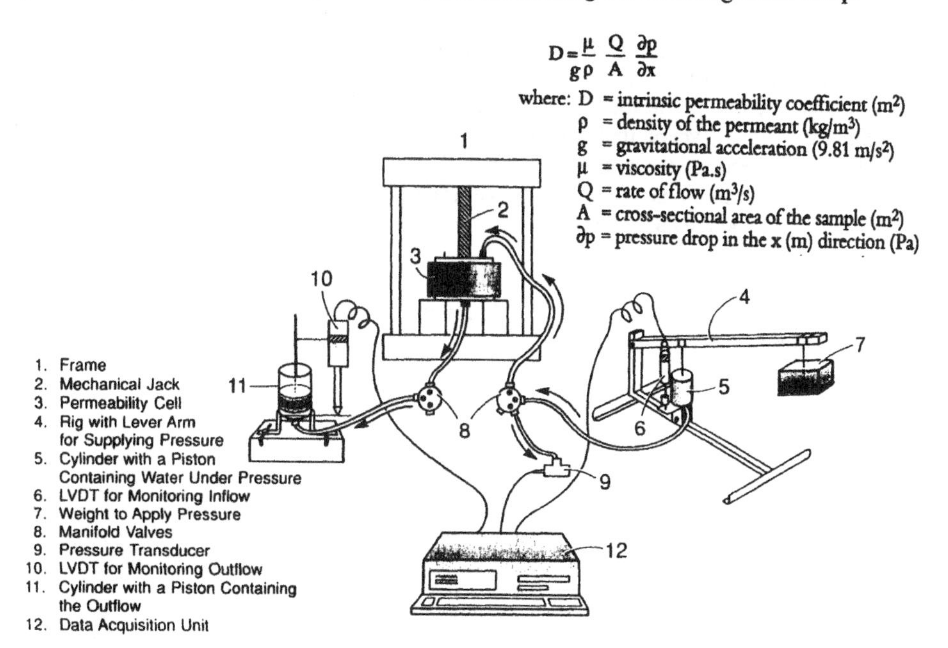

Figure 1. Permeability equipment.

Typical water permeability curves are shown in Figure 2 and summarized in Table2. Based on the water permeability results, the following observations can be made:

1. Under virgin condition, concrete samples exhibit constant permeability (Figure 2a). The intrinsic permeability was in the order of $1\text{x } 10^{-19}\text{m}^2$.
2. Once the concrete microstructure is cracked (in this case through shrinkage drying), the permeability initially increases by 2 orders of magnitude due to the increase in the size of the continuous pore radius created by the interconnected system of micro-cracks. Subsequently, the permeability decreases as shown in Figure 2b and Table2. In some cases, the damage created by shrinkage cracking is almost completely reversed by the sealing process.

Propan-2-ol Permeability Tests

Propan-2-ol permeability tests were conducted on the permeameter that was used for water permeability. Propan-2-ol was used because of its lack of chemical interaction with the cementitios matrix.

Several concrete specimens were oven dried and saturated with water placed in propan-2-ol until constant weight was reached, for propan-2-ol replacement of pore water. Subsequently, permeability tests were conducted in the same manner as the water permeability testing, except that propan-2-ol was used as the permeating fluid. After several days of testing, the permeating fluid was changed to water. These test were conducted in order to assess if purely physical transport was contributing to the reduction in flow during the permeability tests.

The results are shown (Table 3) that self-sealing is altogether absent in the samples tested with propan-2-ol, while subsequent introduction of water resulted in the decrease in permeability.

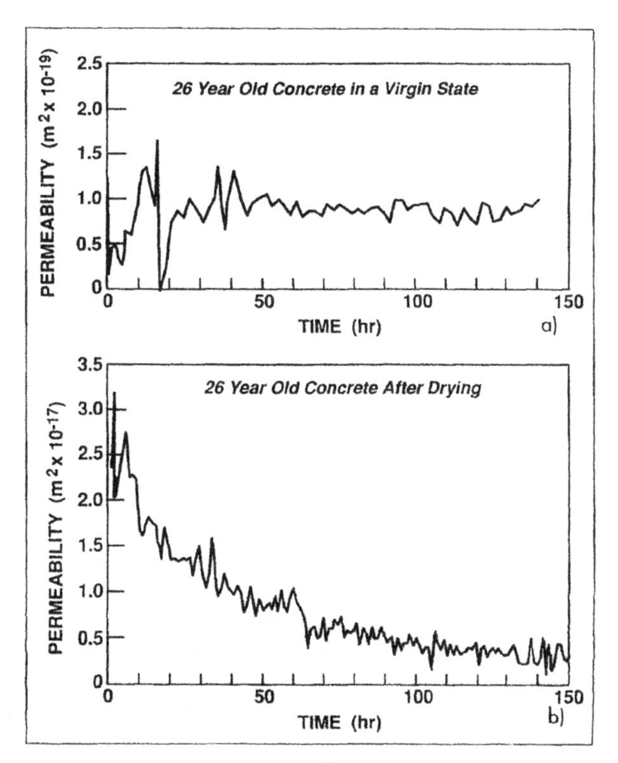

Figure 2. Typical variation of the permeability coefficient during testing (a) 26-year old, never-dried concrete; and (b) the same concrete tested after oven drying at 105°C and resaturation.

26-year old concrete cylinder ID. #	Permeability D x 10^{-19} (m^2) Water	
	1 Virgin	2 D/W*
7-4	0.9	160→1
12-1		300→20
12-2	0.7	260→20
12-3	1	180→20
11-3	0.8	110→8.5
11-4	2	128→21
9-1	0.4	150→2.5
9-2	1.3→0.3	80→6
9-3	0.6	140→1.2

** D/W-oven dried and resaturated with water following AASHTO T267 before permeability testing.*

Table 2. Summary of water permeability results.

Pore Solution Analysis

Supply and effluent water from the permeability apparatus was systematically collected for analysis. Analysis included, chemical analysis using atomic absorption, pH and electrical conductivity. Analysis of the inflow and outflow were conducted before and after drying, in order to determine changes in the pH, electrical conductivity and Ca^{2+}, Na^{+}, K^{+} and Mg^{2+} concentrations as the water percolated through the concrete. The results are given in Table 4, with the top number representing the average readings, and the numbers in the brackets – standard deviation. In several cases, the effluent was collected several times during the test in order to monitor concentration changes as the permeability test progressed. These results are shown in Table 5, and can be summarized as follows:

1. The pH increased as water passed through the specimen. Characteristically, the pH of water at the top of the specimen (on the inflow side) was higher than that of the supply water in the reservoir, thus indicating osmotic mixing of the supply water with pore solution.
2. The pH of the effluent in the virgin specimens was lower than that of oven-dried and re-saturated specimens.
3. The electrical conductivity of pore water was considerably higher for the effluent compared with that of the supply water.
4. Calcium concentration in the effluent increased dramatically after oven drying and re-saturation, thus indicating exposure of previously unexposed soluble calcium to dissolution.
5. The calcium concentration in the effluent decreased as the volume of the water passing through the sample increased.
6. The concentration of Mg^{+} decreased, with Mg^{+} being depleted from the supply water. Mg^{+} in the water possibly reacts with the cement matrix in the following reaction:

$$CH + MS\ (aq) \rightarrow CSH_2 + MH$$

   This reaction involves both calcium hydroxide and magnesium, with the resulting product having molar volume of 65.6cm^2 [6].
7. The changes in Na^{+} and K^{+} concentrations are small. Their concentrations are similar in both in the inflow and outflow water samples, indicating osmotic mixing of these species.

SEM and X-ray analysis Examination

SEM examination was conducted on concrete samples cut into prisms approximately 30x10x5mm in dimension. The small prisms were oven dried at 105°C until constant weight. The dried samples were then fractured mechanically in half. One set of prisms was taken from concrete samples that did not undergo any permeability testing, while the other set of prism was taken from concrete after permeability testing and self-sealing.

26-year old Concrete Cylinder #	Sample ID	Permeability D x 10^{-19} (m^2)	
		D/W → P (oven dried, saturated with water, replaced with propan-2-ol and tested for propan-2-ol permeability)	D/W → P → W (oven dried, saturated with water, replaced with propan-2-ol and tested for water permeability after propan-2-ol permeability)
11	5	34	8 → 1.7
8	1	39	17 → 3
8	2	50	24 → 11
8	3	41	10 → 3

Table 3. Propan-2-ol permeability data.

Concrete W/C	Conditioning	Inflow Outflow	pH	Elec. Cond. (μ 1/Ω)	Na+ (mg/l)	K+ (mg/l)	Ca++ (mg/l)	Mg++ (mg/l)
Supply Water			7.5	300	10	3	30	8
0.9	Virgin	inflow	9.7* (1.6)+	519 (332)	13 (5)	13 (7.5)	36 (24)	4 (3)
		outflow	10.7 (1.8)	1362 (347)	19 (2.3)	10 (1)	64 (41)	.5 (.6)
	Dry/ resaturated	inflow	8 (1.3)	-	34 (13.5)	21 (22)	17 (11)	8 (3.4)
		outflow	12 (.3)	4537 (1,041)	36 (22)	44 (50)	112 (91)	1.35 (.9)

Table 4. Chemical analysis of the inflow and outflow pore solutions

Concrete W/C	Conditioning	Inflow Outflow	Hours of effluent	pH	Na+ (mg/l)	K+ (mg/l)	Ca++ (mg/l)	Mg++ (mg/l)
0.9	virgin	outflow	60	7.3	-	3.73	30.36	8.65
	dry/saturated	outflow	14	12.75	21.45	21.82	195.6	0.43
		outflow	38	12.35	7.63	18.38	286.5	2.18
		outflow	62	11.93	50.0	23.63	48.7	1.2
		outflow	158	11.94	49.9	24.49	47.19	0.9
		inflow	320	8.61	46.5	22.74	17.66	6.8
		outflow	320	11.91	50.9	24.98	48.18	2.1

Table 5. Typical changes in the effluent with the progress of the permeability test.

A drop of water was placed on one side of the fracture surface, while the other side was kept dry. After two hours, the wetted samples were returned to the oven. The fractured samples were then mounted and coated with 200 Å thick layer of gold for examination under the SEM. The Hitachi 540 electron microscope equipped with Link energy dispersive X-ray was used.

SEM images of samples, which were not exposed to self-sealing, showed considerable growth on re-wetting (Figure 3a). Samples, which have already undergone self-sealing did not exhibit significant differences between re-wetted and dry fracture surfaces (Figure 3b). X-ray analyses also confirm SEM visual interpretation. Those samples, which were dried for the first time and were not tested for permeability after drying, show crystallization of ettringite and spongy calcium-silicate crystals only on the fracture surface that was re-wetted. Those samples, which have exhibited self-sealing, do not show differences between dry and re-wetted fracture surfaces.

The SEM, together with X-ray analysis indicate that:

1. the potential for chemical interaction is at its greatest after the first drying,
2. once the self-sealing has occurred, the potential for further chemical interaction is significantly diminished, and
3. specimens which have exhibited self-sealing contain massive CH deposits.

DISCUSSION

The alkalis in the pore solution have long been known to affect permeability testing and results. Powers' [8] permeability tests were conducted on thoroughly leached cement paste samples. Markestad [9] showed that permeability reached a steady state after the alkalis were either diluted by mixing with fresh water or leached from the concrete mix.

Based on the analysis of the supply and effluent water during the permeability testing, it is evident that calcium plays a significant role. Both pH and electrical conductivity indicate that drying process results in a significant increase in the OH^- concentration and overall amount of dissolved species. Similarly, calcium concentration in the effluent increased dramatically after oven drying and re-saturation. In virgin samples, due to long-term storage of the concrete in water, most of the soluble phases exposed to pore water reached equilibrium dissolution stage. Upon oven drying, drying shrinkage caused cracking between phases of varying moduli of elasticity and shrinkage of calcium hydroxide upon drying, thus exposing to the pore water previously unexposed cementitious matrix. Considering that BSE examination showed that after 26 years most of the clinker was gone, except for some remnant ferrite phases, continuing hydration is not responsible for the changes in composition of the pore solution after drying. Therefore, it is assumed, upon drying, dissolution of the newly exposed fracture surfaces was initiated, in order to establish new equilibrium between pore water and newly created pore walls.

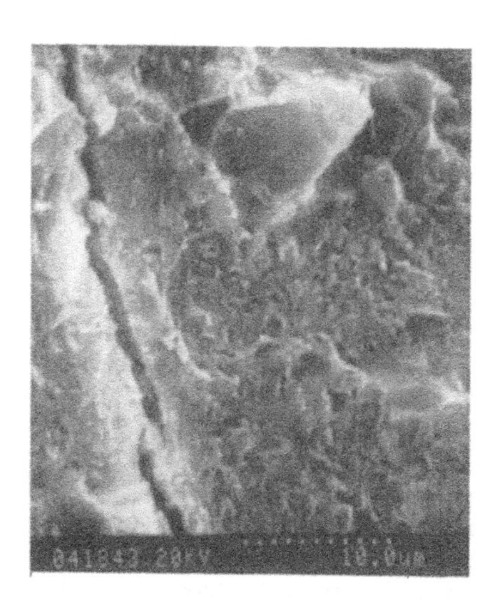

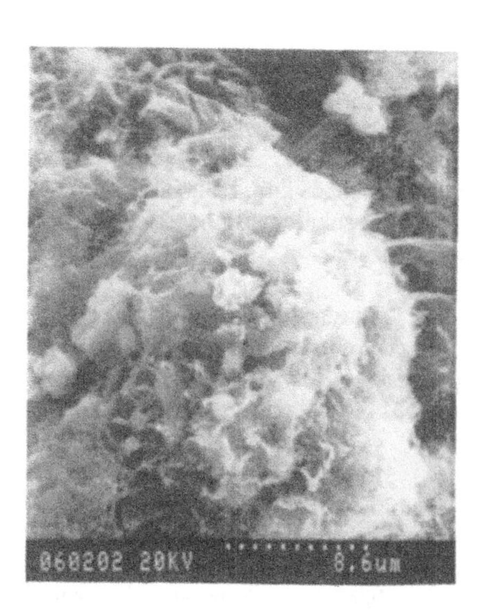

Figure 3a SEM of 26 year-old concrete prior to self-sealing

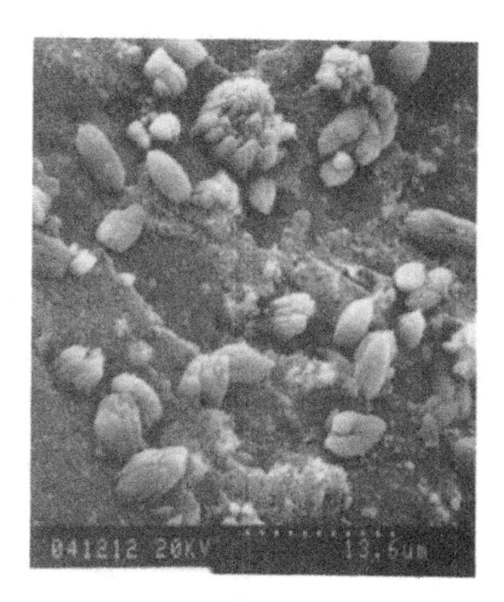

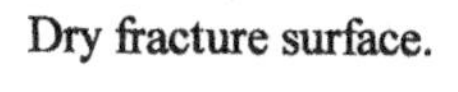

Dry fracture surface. Exposed to water

Figure 3b SEM of 26 year-old concrete after self-sealing

The increased activity on the newly formed fracture surfaces is well demonstrated by the SEM analysis of the fracture surfaces exposed to water (Figure 3). Considerable potential for dissolution and subsequent crystallization exists after oven drying of the concrete samples. X-ray analysis confirmed that most of the newly formed deposits were CH in composition. Similarly, during the water permeability test, the flowing water allows dissolution of the calcium components of the pore matrix and subsequent re-crystallization.

The largest drop in permeability occurred in the first 40 hours of testing, which also corresponded to the highest concentration of calcium in the effluent. As the permeability test continued, the concentration of calcium decreased, as newly available sources of calcium were depleted and the leaching process set in (Table 5).

In order to demonstrate that the self-sealing phenomenon is dependent on the water dissolving of calcium through the pore structure, propan-2-ol permeability tests were conducted. The results in Table 3 show that permeability remained constant when propan-2-ol was used as a permeant, while on re-introduction of water a decrease in permeability was observed, this result also supporting the chemical nature of the self-sealing process.

CONCLUSIONS

The availability of calcium hydroxide in cementitious matrix provides concrete with the ability to heal its "cracking wounds". This healing process is possible not only in large cracks, as in the case of the autogenous healing, but is continuously occurring in the cementitious microstructure in the presence of water.

Shrinkage cracking exposes previously unexposed solute, thus making the self-sealing process more pronounced. Under daily exposure, concrete matrix undergoes similar process of dissolution and recrystallization of the calcium phases. This process, however, is less pronounced as it occurs together with other processes, such as continuing hydration, drying and wetting regimes accompanied by swelling and shrinkage. By a particular choice of samples and conditioning techniques, this study was able to isolate the effects of calcium hydroxide on the permeation characteristics of concrete.

The questions not answered by this study are:

1. *How much* calcium hydroxide is needed for self-healing?
2. Can we produce a blended cement-based concrete containing just enough calcium hydroxide to give us the described benefit?
3. Is self-healing important enough to warrant keeping high levels of calcium hydroxide in concrete? Or are the negative aspects of presence of calcium hydroxide overwhelming?

REFERENCES

1. Hearn, N. 1996. "Comparison of Water and Propan-2-ol Permeability in Mortar Specimens", *Advances in Cement Research*, Vol. 8, No. 3, pp. 81-86.

2. Garboczi, E.J., and Bentz, D.P., 1989, "Fundamental Computer Simulation Models for Cement-Based Materials", in *Materials Science of Concrete* II, (eds., J. Skalny and S.Mindess), Amer. Ceramic Soc., pp/ 249-277.

3. Lauer, K.R. and Slate, F.O., 1956, "Autogenous Healing of Cement Paste", *J. American Concrete Institute* – Proc., pp. 1083-1097.

4. Clear, C.A., 1985, "The Effect of Autogenous Healing upon Leakage of Water Through Cracks in Concrete", Technical Report 559.

5. Wagner, E.F., 1974, "Autogenous Healing of Cracks in Cement-mortar Linings for Grey-iron and Ductile-iron Water Pipes", J. Am. Water Works Assoc., Vol. 66, pp.358-360.

6. Mindess, S., Young, J.F., 1981, *Concrete*, Prince-Hall Inc.

7. N. Hearn, R.H. Mills, 1991, "Simple Permeameter for Water or Gas Flow", *Cement and Concrete Research*, Vol. 21, pp. 257-261.

8. Powers, T.C., Copeland, L.E., Hayes, J.C. and Mann, H.M., 1954, "Permeability of Portland Cement Pastes", *J. Amer. Concrete Inst.* Vol. 51, pp.258-298.

9. Markestad, A., 1977, "An Investigation of Concrete in Regard to Permeability Problems and Factors Influencing the Results of Permeability Tests", C&CRI, Norwegian Inst. Of Tech., STF 65A 77027

Ready for lunch: Sidney Mindess, Andrew Boyd,
Sidney Diamond, and John Figg

Fred Kinney, David Myers, and Natalyia Hearn

INVESTIGATIONS BY ENVIRONMENTAL SEM: $Ca(OH)_2$ DURING HYDRATION AND DURING CARBONATION

J. Stark, U. Frohburg and B. Möser
Bauhaus-Universität Weimar
(FIB)
Coudraystraße 11
D-99421 Weimar, Germany

ABSTRACT

An Environmental Scanning Electron Microscope (ESEM) was used to explore early stages of cement hydration. Because the observations were performed at low vacuum on uncoated samples, novel features on the early hydration products, including C-S-H phases and portlandite, were observed.

Differences exist between the $CaCO_3$ formed by carbonation of $Ca(OH)_2$ and C-S-H phases produced during hydration of ordinary portland cement and that formed during the hydration of cement rich in ground granulated blast furnace slag. It will be shown that of $Ca(OH)_2$ and C-S-H phases in cements rich in ground granulated blast furnace slag carbonate very fast by forming metastable modifications vaterite and aragonite.

During freezing and thawing, vaterite and aragonite transform into poorly-crystallized calcite, which then leads to concrete surface scaling that is much more severe than with other types of concrete.

CONCERNING EXPERIMENTS USING THE ENVIRONMENTAL SCANNING ELECTRON MICROSCOPE

The F.A.Finger Institute at the Bauhaus University is fortunate enough to use the X130 ESEM-FEG, a Philips-made microscope with integrated EDX equipment for its investigations. The ESEM wet mode setting employs water vapour as the image creating medium thereby not only delaying but actively preventing dehydration of the samples examined without altering their structure at all. Whilst being examined the sample may be moisturised through condensation in the supersaturated water vapour atmosphere or by injecting water from the microinjector. For there is hardly any contamination of the surface at all, neither the high-resolution scan nor the microanalysis is distorted by this phenomenon. In addition, no conductive coating of any kind is required.

If it is verified in comparative tests that, within the field of definition, no structural alteration of the sample to be examined is caused by the vacuum inside the SEM, high definition images from non-conductive, non-coated and delicate preparations may be obtained using a field emission SEM set within the low voltage range (mode: low-kV/high-vac).

A reduced stimulant voltage enhances the topographical contrast, thus even samples containing low density phases such as ettringite may be shown in high definition images.

Another advantage of stimulating within the low voltage range is the chance to obtain analytical data within a nanometer spectrum, by means of electron ray microanalysis using an X-ray spectrometer (spectroscope) of ultra-high energy definition and very high sensitivity.

Figure 1 illustrates the stimulant quantity of the electron beam as determined by Monte Carlo simulation with 4'000 electrons at 15 and 2.5 kV of stimulant voltage for a 1.13 nm tobermorit crystal [$Ca_5(Si_6 O_{18} H_2)4 H_2O$].

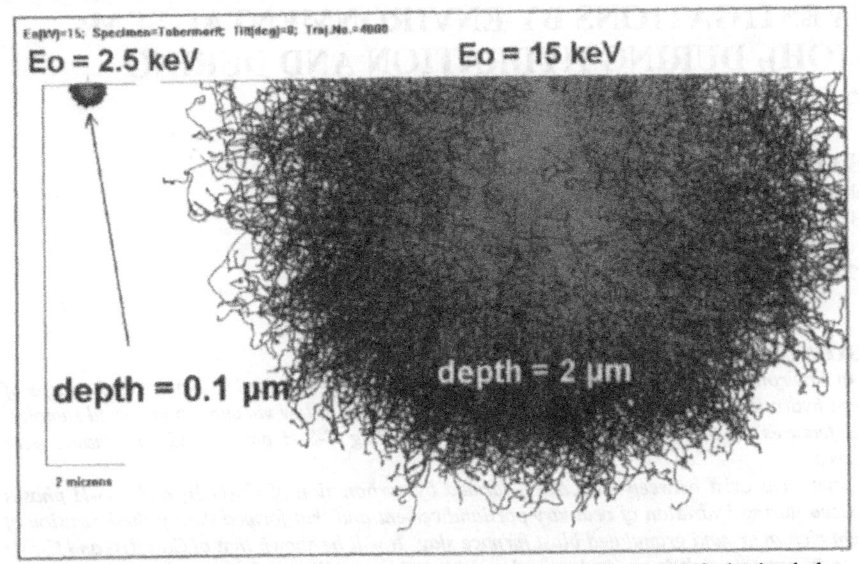

Figure 1. Stimulant quantity of the electron beam as determined by Monte Carlo simulation with 4'000 electrons at 15 kV (right) and 2.5 kV (left) of stimulant voltage for a 1.13 nm tobermorit crystal [$Ca_5(Si_6 O_{18} H_2)4 H_2O$]

As can be clearly seen the standard procedure of high voltage stimulation effects a quantity of some µm³. Thus, some sort of "conglomerate analysis" is obtained for phases of low density (C-S-H at about 2.2 g/cm³) and less than 1µm in size. In comparison, the low stimulant voltage yields a spatial resolution between 50 and 100nm which relates to the approximate thickness of a single C-S-H phase (see figure 2 and 3).

The figures shown in this paper were generated using the ESEM-WET mode or the low-kV/high-vac mode settings.

FORMATION AND MORPHOLOGY OF PORTLANDITE

The hydration of C_3S and C_2S (in ordinary portland cement/OPC) due to the formation of C-S-H phases and $Ca(OH)_2$ (CH, portlandite). The morphology of these phases may be very different, for instance in dependence on the degree of hydration. In general, portlandite occurs in aggregations (figure 4-6) but also single crystals of portlandite have been recognised in bundles of C-S-H phases (see figure 2).

Figure 2. Single crystals of portlandite in bundles of C-S-H phases (hydrated C_3S)

A contribution of portlandite to the steadfastness (strength) of concrete seems unlikely as the comparison in sizes between C-S-H and portlandite shows (figure 2 and 3).

Figure 3. Portlandite and C-S-H phases in hydrated C_2S paste

The structure of the portlandite aggregates in hydrated OPC is formed by stacks of portlandite crystals 20 - 200 nm thick. The surface of each crystal is fairly even, the surface of the whole structure is rather flaky (figure 4 and 5).

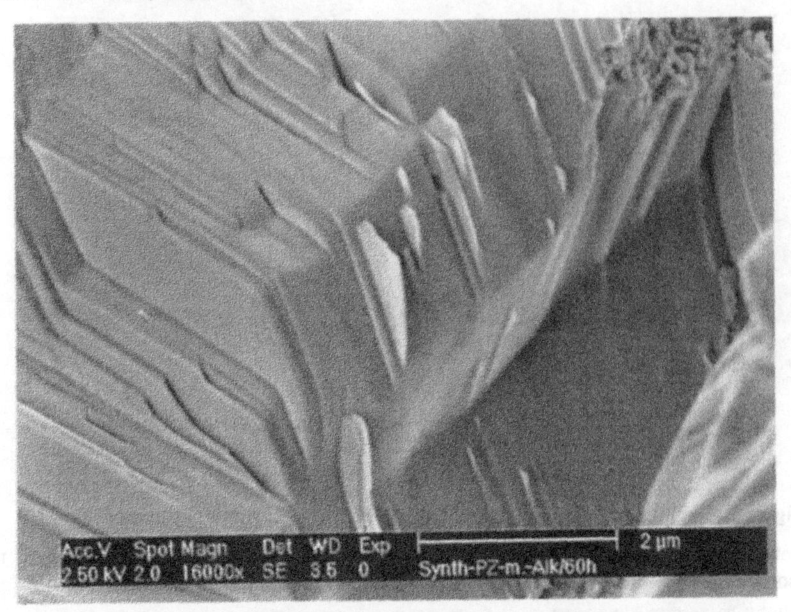

Figure 4. Portlandite aggregate, stacks of portlandite crystals 20 - 200 nm thick

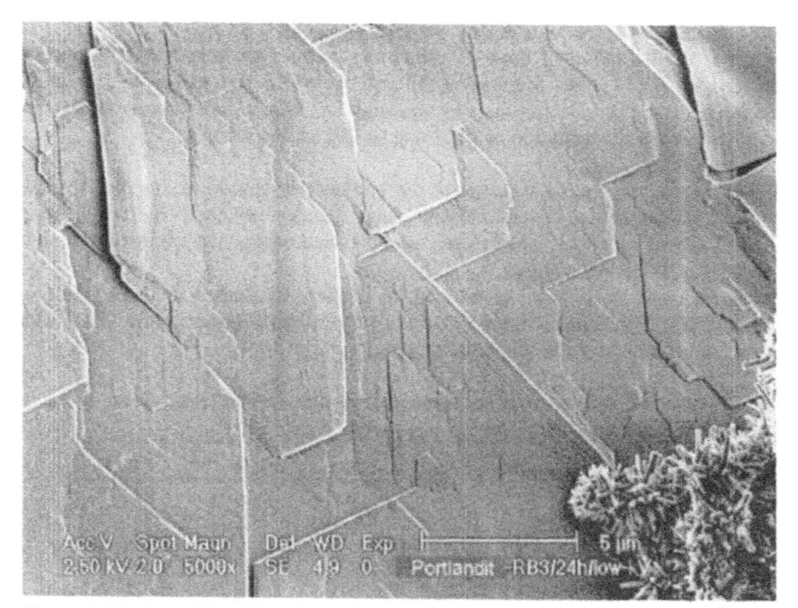

Figure 5. Flaky surface of portlandite aggregates

Figure 6 illustrates the formation of portlandite on the interface of quartz aggregate and matrix.

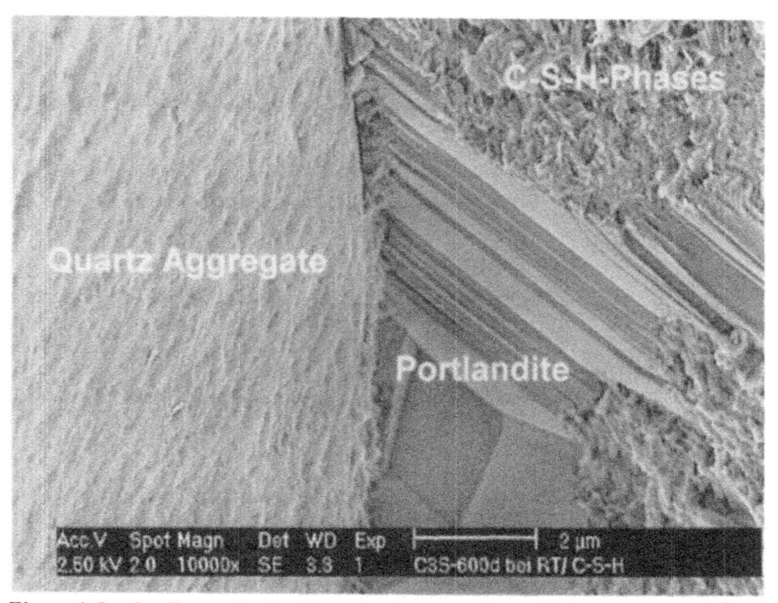

Figure 6. Portlandite on the interface of quartz aggregate and matrix

HYDRATION OF CEMENT CONTAINING BLAST FURNACE SLAG

It is known that the hydration of cements containing ground granulated blast furnace slag (further abbreviated to 'slag') results in a smaller amount of portlandite (calcium hydroxide, CH) [1]. The reason is the dispersing effect of the slag in connection with a lower content of the clincer phases C_3S and C_2S. During the hydration of 'pure' slag no CH will be formed.

Table I shows, exemplary, the dependence of the portlandite content on the slag content of cement. For these investigations certain cements were custom produced in the laboratory, containing slag from 15 up to 75%. Cement pastes were prepared using a water/cement ratio of 0.5 and cured for 7 days at 20°C/100% relative humidity (r.h.).

The portlandite contents have been determined by differential thermal analysis (DTA) and thermogravimetry (TG). The storage chosen allowed for carbonation to start. The calcium carbonate contents measured are listed below in brackets.

Table I. Portlandite content in cement stone in dependence of the slag content of the cement

slag content [%]	$Ca(OH)_2$ %]	($CaCO_3$) [%]
15	12,3	(9,3)
30	10,2	(9,1)
45	8,1	(7,8)
60	5,1	(7,4)
75	3,0	(6,6)

The XRD patterns belonging to these cement pastes are shown in the following figure 7.

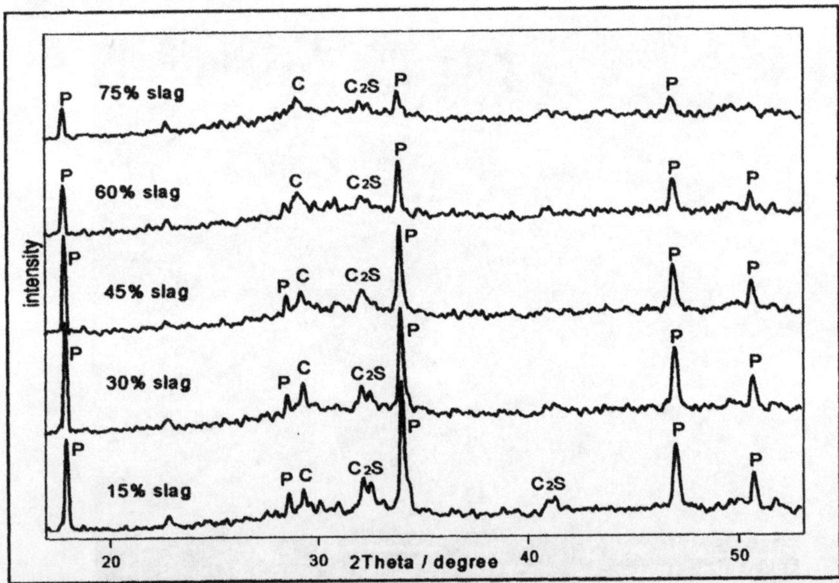

Figure 7. XRD patterns of cement pastes with different slag content (stored at 20°C/100% r.h. for 7 days), P - portlandite, C - calcite

A large amount of portlandite with large crystals could be found in portland cement paste and also in cement pastes with a low slag content after 7 days of hydration. The products of hydration in cement pastes rich in slag are rather 'gel-like' and it is more difficult to detect portlandite. The higher the slag content, the more difficult it is to identify well crystallized portlandite.

This should be clarified with a few frames, taken with ESEM and/or SEM. The figures (8-11) show portlandite in hydrated portland cement paste and in blast furnace cement paste with 53% slag and 75% slag, respectively. Clearly, the hydration of C-S-H (calcium silicate hydrate) phases progresses differently as well, in connection with differences in the morphology.

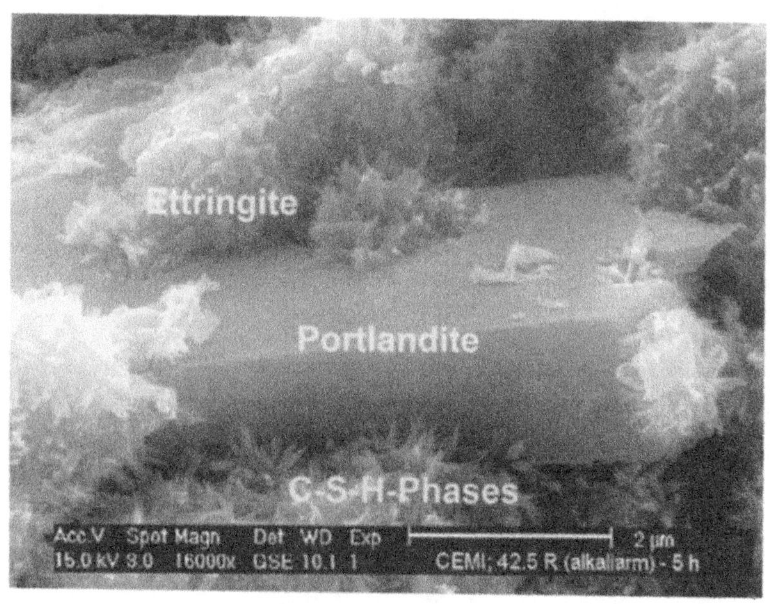

Figure 8. Portlandite ettringite and C-S-H phases in hydrated portland cement paste after 5h

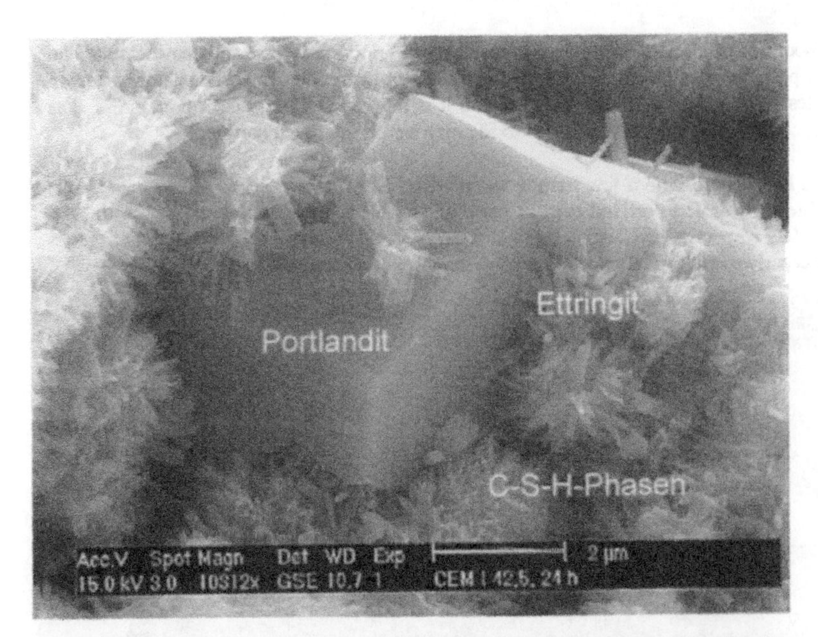

Figure 9. Portlandite and C-S-H phases in hydrated portland cement paste after 24h

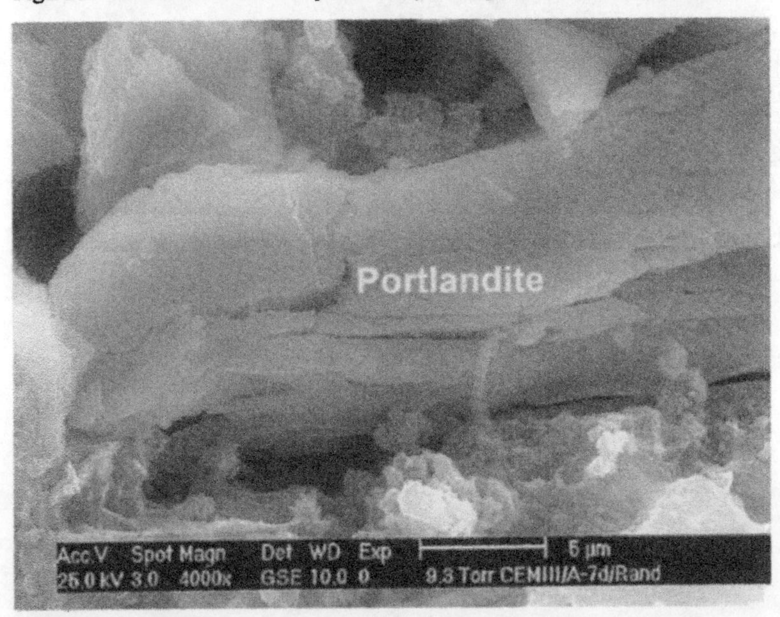

Figure 10. Portlandite and C-S-H phases in hydrated blast furnace cement paste (53% slag) after 7d

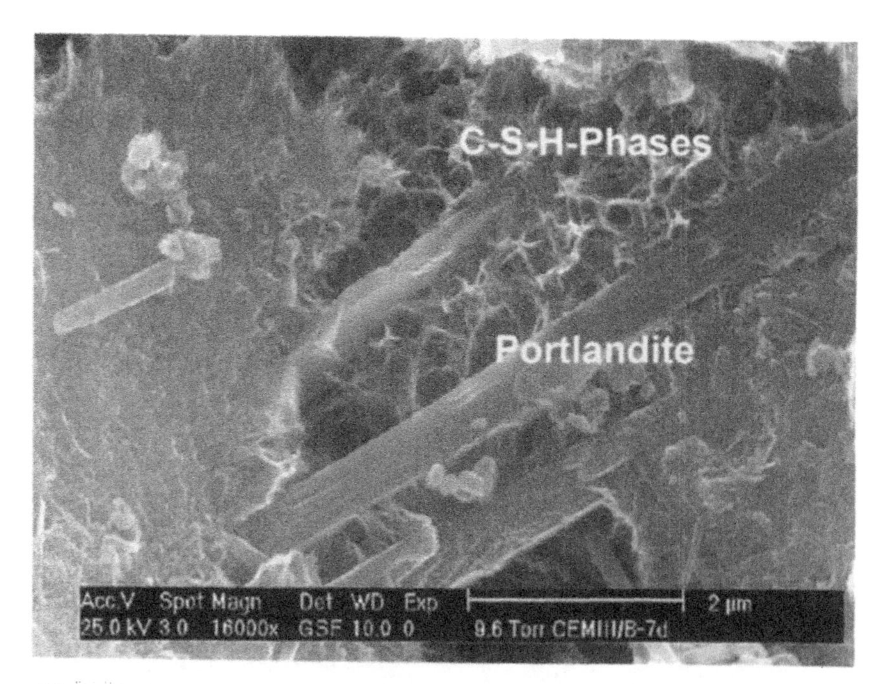

Figure 11. Portlandite and C-S-H phases in hydrated blast furnace cement paste (75% slag) after 7d

CARBONATION OF CEMENT CONTAINING BLAST FURNACE SLAG

In the process of carbonation of the hydrated phases in hardened cement paste, besides calcite as the metastable polymorphs of calcium carbonate, aragonite and vaterite are formed [2-7]. These modifications are already detectable in cements with low slag content under certain conditions. If a slag content of about 50% is exceeded, the calcium carbonate formed consists of such a high amount of metastable modifications, that a substantial influence on the durability of such concretes must be stated. The major characteristics of the three modifications of calcium carbonate as well as the ones of portlandite are listed in table II.

Table II. Characteristics of crystalline calcium hydroxide and calcium carbonate [8-10]

name	formula	crystal system	habits morphology	density [g/cm³]	
portlandite	$Ca(OH)_2$	trigonal	tabular, prismatic...	2,23	
calcite	$CaCO_3$	trigonal	tabular, prismatic, acicular, rhombohedral...	2,71	most stable
aragonite	$CaCO_3$	orthorhombic	columnar, fibrous, acicular..., (mostly untwinned: → pseudo hexagonal)	2,95	less stable
vaterite	$CaCO_3$	hexagonal	spherolitic, tabular...	2,54	least stable

As table II shows, the morphology of the three calcium carbonate modifications may be very different. Therefore it is partly very difficult to identify these modifications. Results from additional investigation such as X-ray analysis (XRD) have to be compered to the ESEM data. Yet, even by means of XRD small amounts of these metastable modifications are hard to be identified because of there crystalline structure.

The following frames (figure 12 and 13) are to illustrate the change of modification of $CaCO_3$ taking the transformation of vaterite to calcite as an example. For these investigations 'pure' i.e. synthetically produced $CaCO_3$ phases were used.

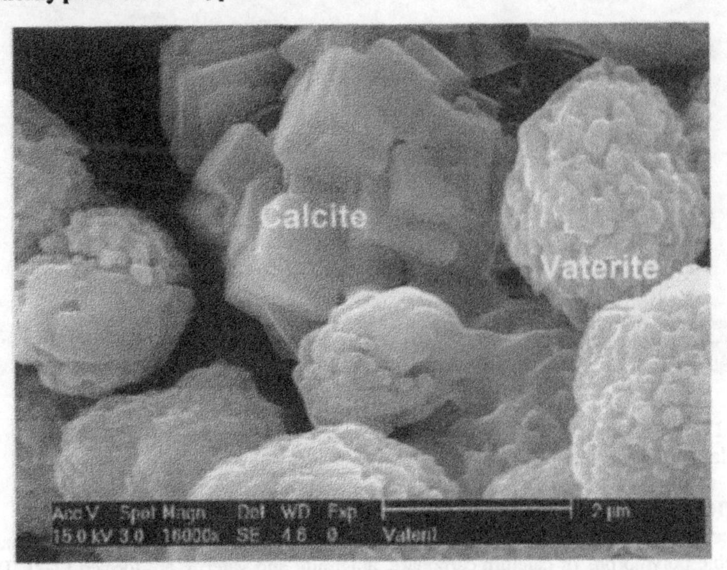

Figure 12. Change of modification of $CaCO_3$, transformation of the less stable, hexagonal vaterite

Figure 13. Modification of $CaCO_3$, stable, trigonal calcite

In order to specify the influence of carbonation the cement pastes with slag contents from 15 up to 75%, the original storage of 7 days at 20°C and 100% r.h. was prolonged by 21 days at 20°C and 65% r.h. (in the laboratory) according to CDF procedure [11]. The existent phases were determined by XRD (figure 14).

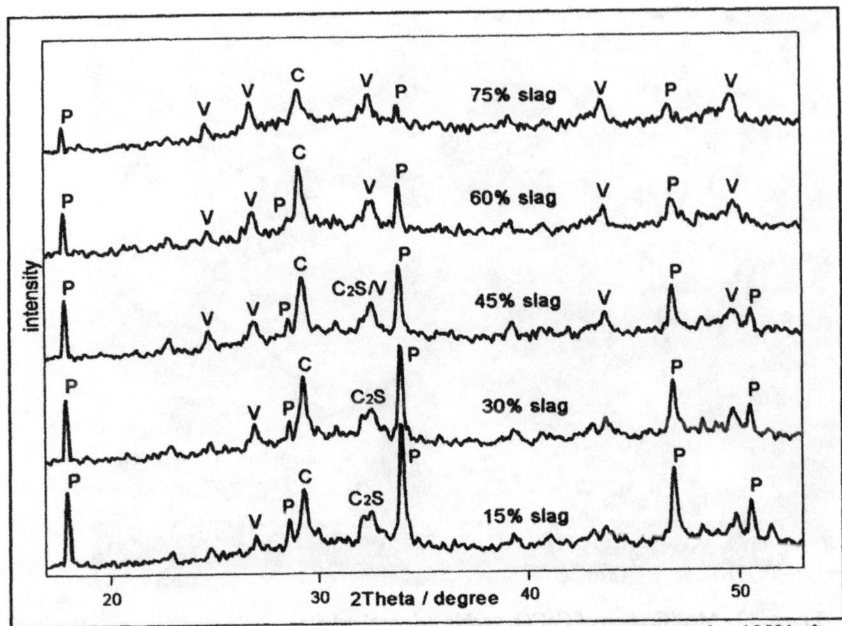

Figure 14. XRD patterns of cement pastes with different slag content, (stored at 100% r.h. for 7 days and at 20°C and 65% r.h. for 21 days), P - portlandite, C - calcite, V - vaterite

It is clearly to be seen, that the proportion of portlandite was reduced in all samples. But even in the sample containing 75% slag, portlandite is still detectable. As a product of hydration in all of these samples, in addition to calcite, vaterite was formed. The content of vaterite increases with higher slag content, as opposed to carbonated portland cement stone, where only calcite was identified (see figure 19).

For further investigation two of-the-shelf cements containing slag were used, i.e. CEM III/A with 53% slag and CEM III/B with 75% slag. A water/cement ratio of 0.4 was fixed for mixing the cement paste, which after one day of hardening in the mould was stored for 6 days under water.

Already after 7 days of hydration carbonation had started as the following frames (figure 15, 16) show. By means of XRD calcite was detected in the hardened cement paste.

Figure 15. Carbonation of blast furnace cement paste (53% slag) after 7d of hydration, needle-like $CaCO_3$ crystals on a grain of slag, (modifications: calcite and aragonite?)

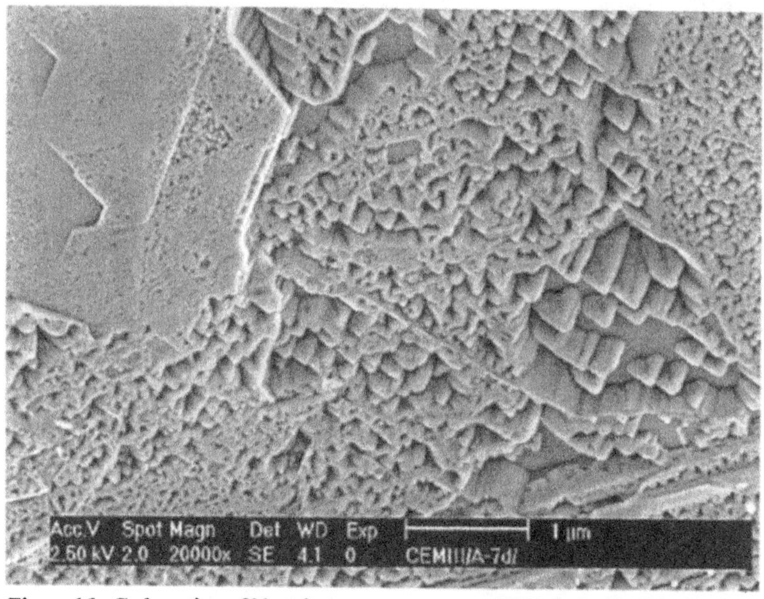

Figure 16. Carbonation of blast furnace cement paste (53% slag) after 7d of hydration, the formerly even surface of the CH crystals now shows highly varied formations of $CaCO_3$

After the afore mentioned 7 days, the storage of these samples was continued in 1% CO_2 atmosphere at 20°C and 65% r.h. for 6 days for further carbonation. The samples (fresh fracture surfaces) were examined with ESEM/SEM, the frames (figure 17, 18) shown represent the CEM III/A with 53% slag as an example.

Figure 19 is a juxtaposition of the XRD patterns of the two afore mentioned blast furnace cement pastes with the one of pure portland cement paste, stored accordingly.

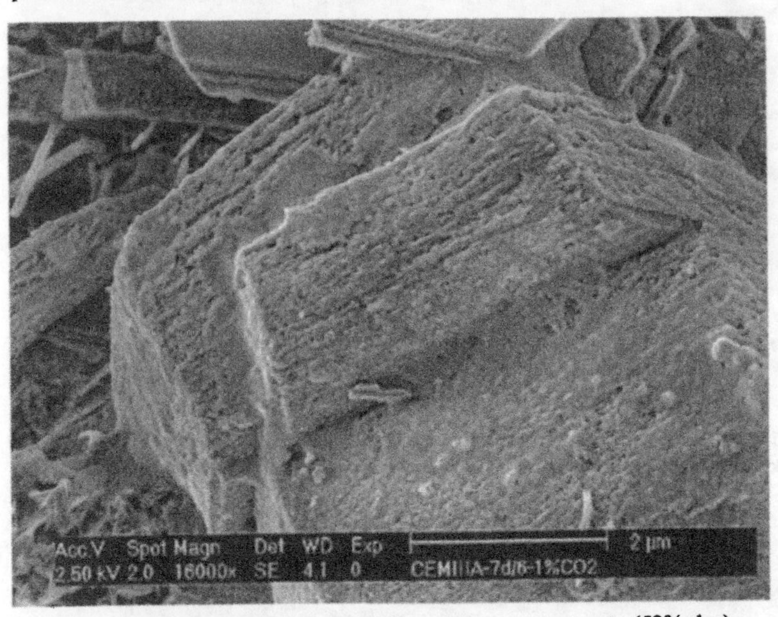

Figure 17. Carbonation of CH crystals in blast furnace cement paste (53% slag), $CaCO_3$ (modifications: calcite and vaterite ?)

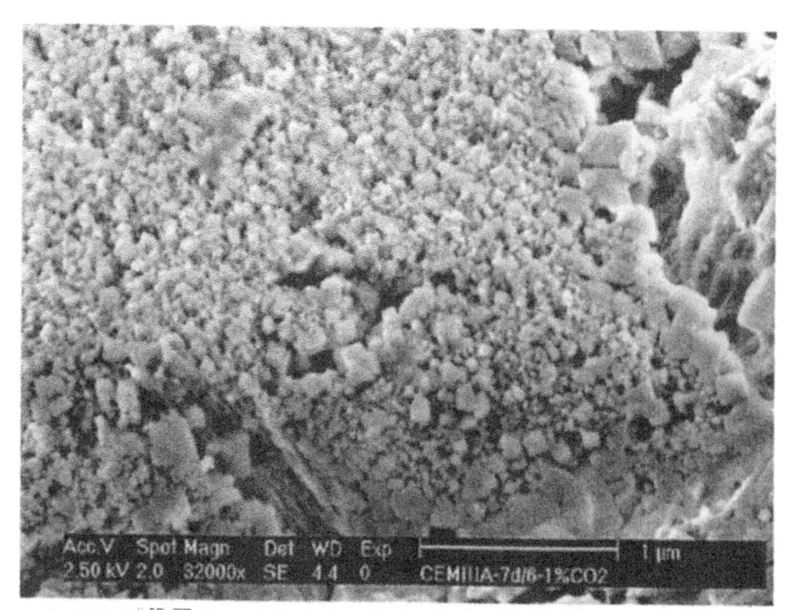

Figure 18. Carbonation of blast furnace cement paste (53% slag), carbonated layer, sublayer not visible, $CaCO_3$ (modifications: calcite and vaterite)

Carbonation progressed strongly. Some of the sample are covered with a layer of carbonation products to such an extent that the phases and/or original substances below are no more recognizable.

It may be concluded, that in cement rich in slag portlandite as well as the C-S-H phases are subjected to carbonation. Yet, calcite, vaterite and aragonite as the substances likely to be formed by carbonation cannot be clearly identified.

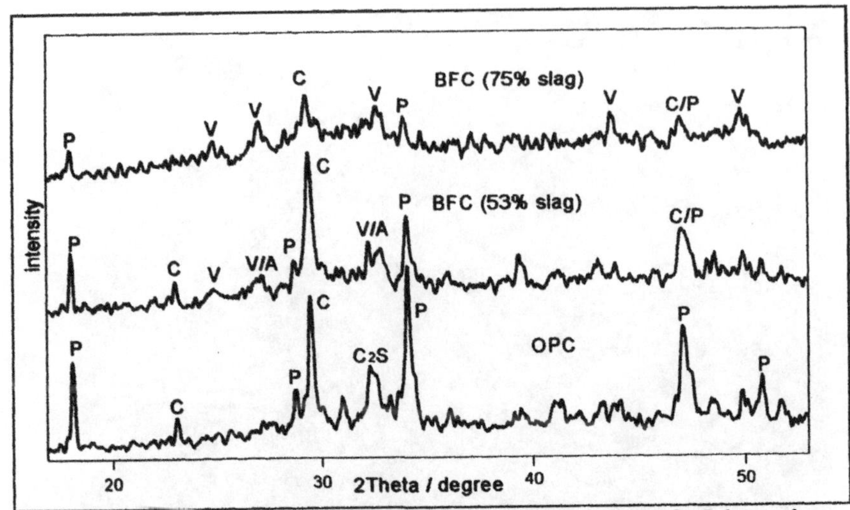

Figure 19. XRD patterns of different cement pastes (stored under water for 7 days and 6 days at 1% CO_2/20°C/65% r.h.), P - portlandite, C - calcite, V - vaterite, A - aragonite

The XRD patterns illustrate, that in the course of carbonation of cements rich in slag in addition to calcite the metastable modification of calcium carbonate vaterite emerged. In the sample originally containing 53% slag even some aragonite may have formed, but if only in hardly detectable quantities.

An array of factors, such as pH-values, temperature, foreign ions additives, and their interaction have an influence on the formation, the stability and the rate of phase transformation of the metastable modification of calcium carbonate [2,3,7-10]. It must be assumed that vaterite (aragonite) is formed not only through the carbonation of C-S-H phases but also through the carbonation of portlandite under certain conditions, such as pore solution of a certain quality. Yet, no final conclusion could be established in this investigation.

Influence of different parameters on the formation of the modifications of $CaCO_3$

Additional tests were performed in order to determine the influence of certain singular parameters.

First, the impediment of calcite crystallisation by Mg^{2+} ions in the pore solution of the blast furnace cements rich in MgO has been investigated. Six synthetically produced slags (0,3 - 7,8% MgO content) and 10 customary blast furnace slags (3,8 - 10,2% MgO content) were used.

The mixtures of 95% slag and 5% calcium hydroxide have been chemically activated by saturated $Ca(OH)_2$ solution and than stored for 7days at 100% r.h. and 20°C and for 21days at 65% r.h. and 20°C.

For all examined slag mixtures carbonation leads to the formation of the metastable $CaCO_3$ aragonite and/or vaterite among the stable calcite. A dependence of the carbonation products on the MgO content could not be found. In figure 20 the XRD patterns of slag pastes from slag with MgO contents of 0.3%, 4.5%, 7.3% and 10.2% are shown.

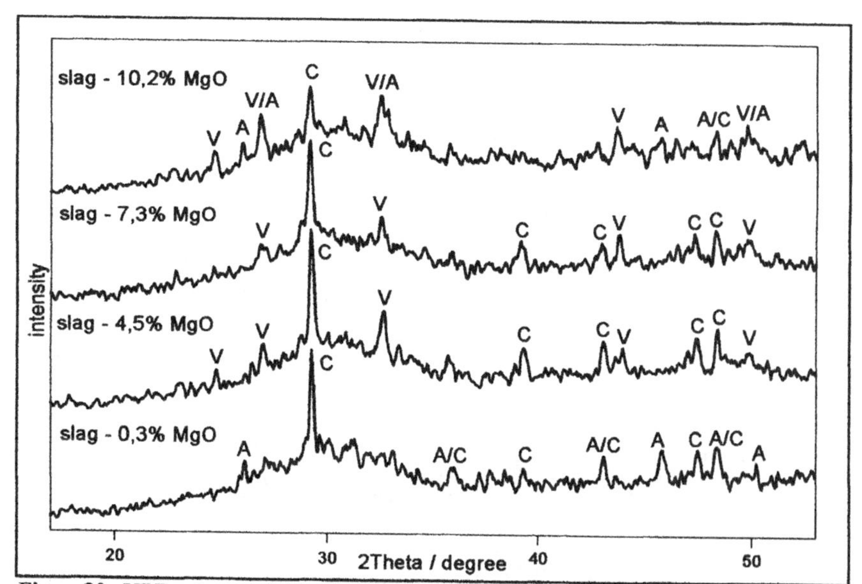

Figure 20. XRD patterns of slag pastes from slag with MgO contents of 0.3%, 4.5%, 7.3% and 10.2% (stored at 100% r.h. for 7 days and at 20°C and 65% r.h. for 21 days), C - calcite, V - vaterite, A - aragonite

To determine whether or not the chemical composition of the slags, here the MgO content, may influence the frost de-icing salt resistance laboratory-produced blast-furnace cements (slag content 75%) were used. Three customary slags were chosen, which differ in MgO content as well as in the amount of formed $CaCO_3$ modifications: the slags with 4.5% MgO, with 7.3% MgO and with 10.2% MgO (figure 20).

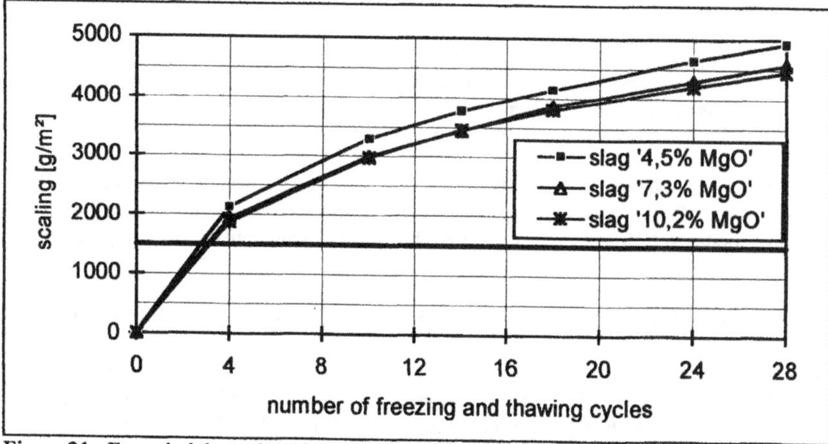

Figure 21. Frost de-icing salt resistance of blast furnace cement concrete from slag with different MgO contents (CDF test [11]; cement content 370 kg/m³, w/c = 0.5)

The three investigated concretes showed nearly the same amount of scaling as well as the same course of scaling (CDF test [11]/figure 21). The amounts of scaled material were ranging between 4500 and 5000 g/m^2, which is usual for blast furnace cement concretes with such a high slag content.

The investigations show that an easy influence of the MgO content of the blast-furnace slag on the formation of the modifications of calcium carbonate does not exist.

Secondly, the influence of the pH-value was investigated. Blast furnace slag pastes have been produced by using different KOH solutions with pH-values ranging from 10 up to 14. Also in this case, an open dependence of the carbonation products on the pH-value of the pore solution could not be detected. All carbonated samples contain calcite, aragonite and vaterite. The content of vaterite was highest in the mixture with a pH-value of 14.

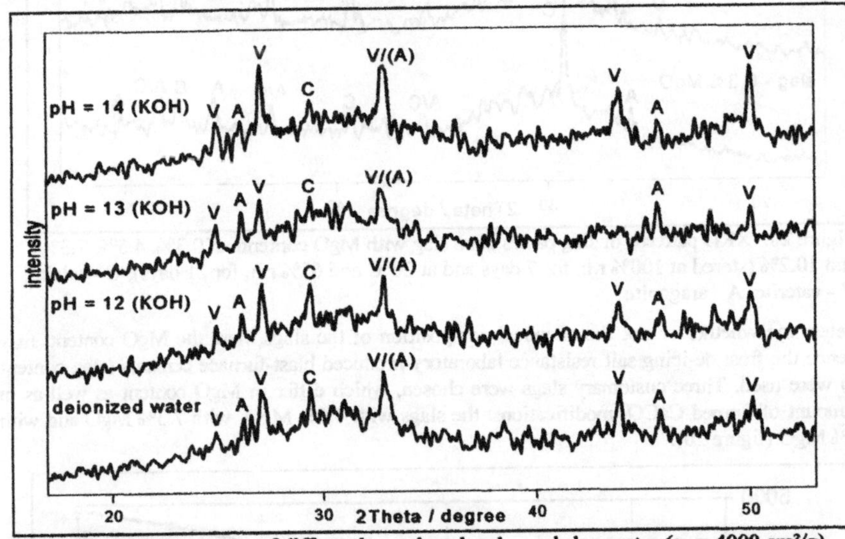

Figure 22. XRD patterns of differently produced and cured slag pastes (a_0 = 4000 cm²/g) (stored under water/solution for 28 days and 28 days at 1% CO_2/20°C/65% r.h.), P - portlandite, C - calcite, V - vaterite, A - aragonite

Figure 22 shows, exemplary, the XRD patterns of slag pastes produced with KOH solutions with pH-values of 12, 13, and 14, respectively, and with deionized water as comparison. The sample stored under solution and/or water for 28 days and afterwards for 28 days in 1% CO_2 atmosphere at 20°C and 65% r.h..

INFLUENCE OF CARBONATION ON THE DURABILITY OF BLAST FURNACE CEMENT CONCRETE RICH IN SLAG

The carbonation of concretes made from cement rich in slag has an effect on their durability [1,5,6,12]. Especially the frost de-icing salt resistance is clearly reduced. Even entraining artificial air voids does not improve their performance significantly (figure 23, 24) [13]. This was observed for concretes with amounts of slag exceeding 50% in their original cement, coinciding with the formation of clearly measurable quantities of vaterite(/aragonite)[5,6].

The frost de-icing salt resistance of all investigated concretes was determined by means of CDF test [11]. After demoulding the specimens were stored 6 days under water and afterwards 21 days at 20°C and 65% r.h., followed by one week of capillary suction (3% NaCl solution). The actual test comprises 28 cycles of freezing and thawing (T_{min} = -20°C and T_{max} = +20°C) within 14 days.

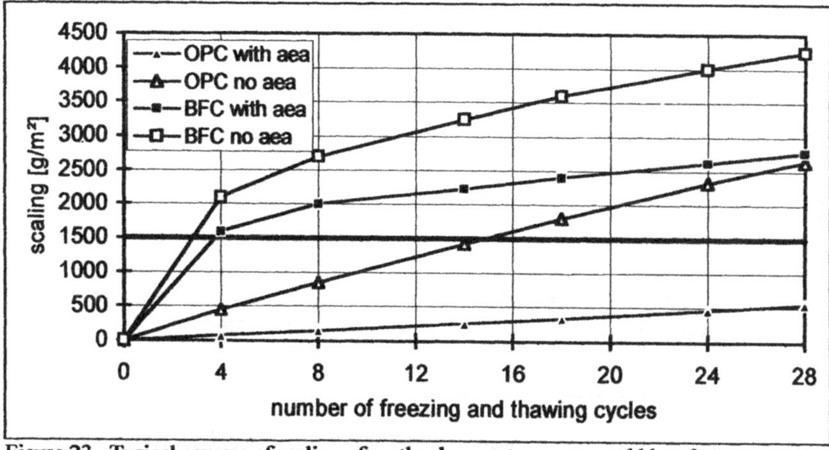

Figure 23. Typical courses of scaling of portland cement concrete and blast furnace cement concrete with and without air-entraining agents (a.e.a.) under frost de-icing salt attack (CDF test [11]; cement content = 350 kg/m³, w/c = 0.5) [5,6]

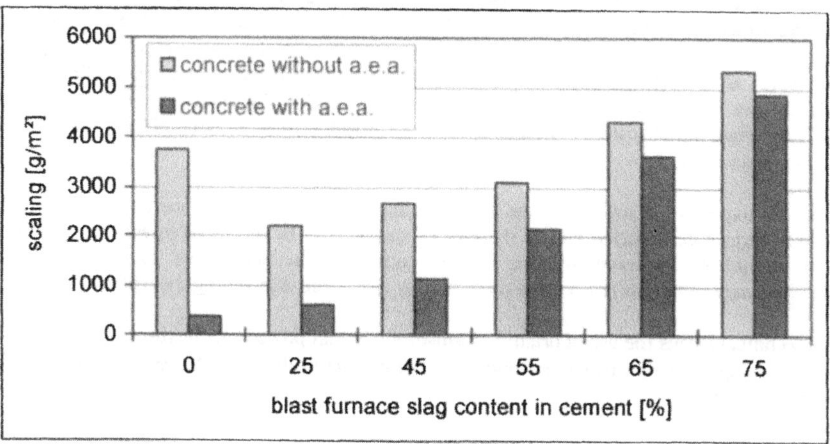

Figure 24. Influence of air-entraining agent (a.e.a.) on the amount of scaling of concretes with cements with different slag content (CDF test [11], acceptance criterion: amount of scaled material ≤ 1500 g/m²; cement content = 350 kg/m³, w/c = 0.5) [5,6]

On one hand, the reduced frost de-icing salt resistance can be explained by the emergence of a coarser microstructure. In concrete from cement rich in slag portlandite as well as the C-S-H phases are subjected to carbonation. The carbonation of C-S-H phases leads to the formation of amorphous silica. This is linked to an increase in capillary porosity and a decrease in gel porosity (figure 25) [5,6,12].

The fine mortar matrix of the concretes were determined by means of mercury high-pressure porosimetry.

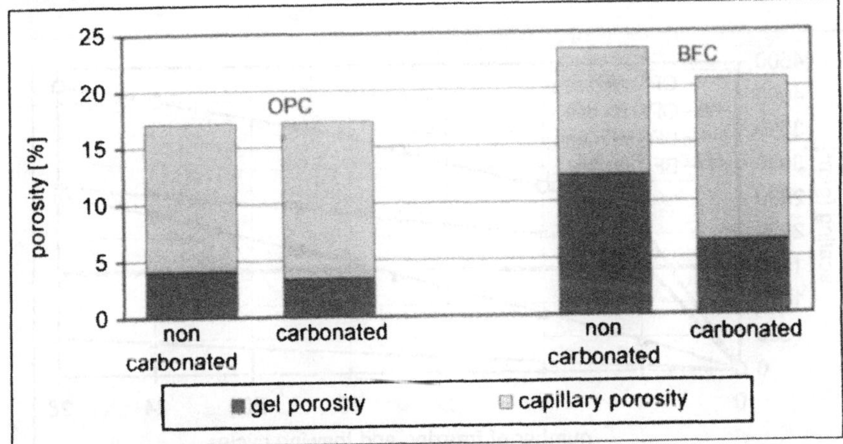

Figure 25. Effect of carbonation on microstructure of fine mortar matrix of portland cement concrete and blast-furnace cement concrete [5,6]

On the other hand, the carbonated layer itself is 'unstable'. The metastable modifications of calcium carbonate are easily soluble in chloride solution (deicing substance). Badly crystallized calcite is formed. The process of dissolution and re-formation during the cycles of freezing and thawing leads to a destabilization of the structure in connection with a rapid scaling of the carbonated layer [5,6]. This explains the typical course of scaling for concretes from cement rich in slag. The strong initial scaling (carbonated fraction) is followed by a period of lesser intensity in scaling (see figure 23).

There are two ways of improving the frost de-icing salt resistance. On one hand, it would be desirable to reduce carbonation and/or the rate of carbonation by technical means. On the other hand, it should be explored whether the formation of the metastable calcium carbonate modifications may be minimized or even prevented by alterations concerning the materials used.

As a technical means the use of draining formwork material proved successful.
The way draining formwork material works, i. e. the change in the water-to-cement ratio shall be illustrated in figure 26.

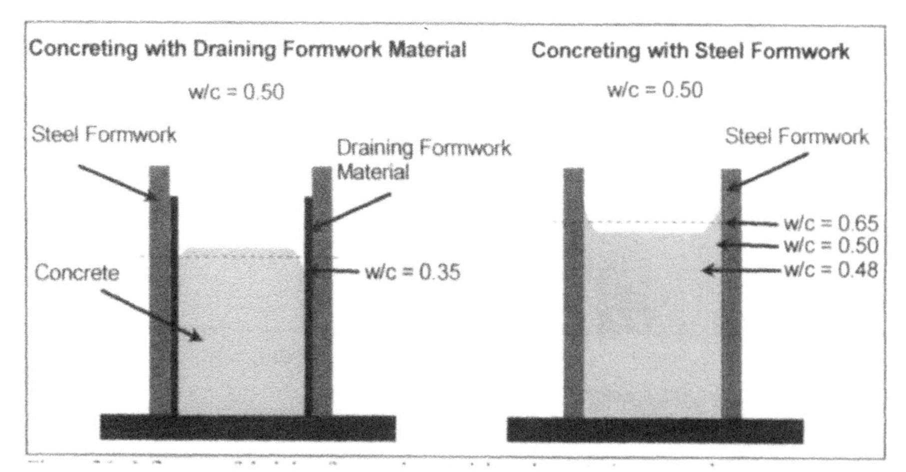

Figure 26. Influence of draining formwork material on the water/cement ratio

In blast furnace cement concrete the microstructure in the surface layer became definitely denser and the depth of carbonation of the concrete could be reduced extremely. As figure 27 shows, the frost de-icing salt resistance of blast furnace cement concrete was increased considerably. The strong initial scaling is prevented and the total amount of scaling is clearly below the acceptance criterion according to CDF.

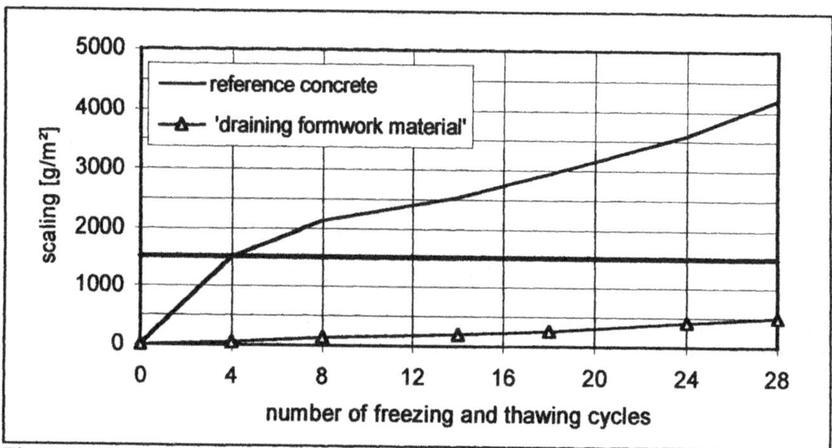

Figure 27. Influence of draining formwork material on frost de-icing salt resistance of blast furnace cement concrete (CDF test [11]; cement content = 350 kg/m^3, w/c = 0.5) [5,6]

Furthermore, the effect of a curing agent, an aliphatic paraffin-wax emulsion, has been investigated. As figure 28 shows carbonation may be reduced clearly. So, this method as well enables the production of blast furnace cement concrete rich in slag with high frost de-icing salt resistance (figure 29).

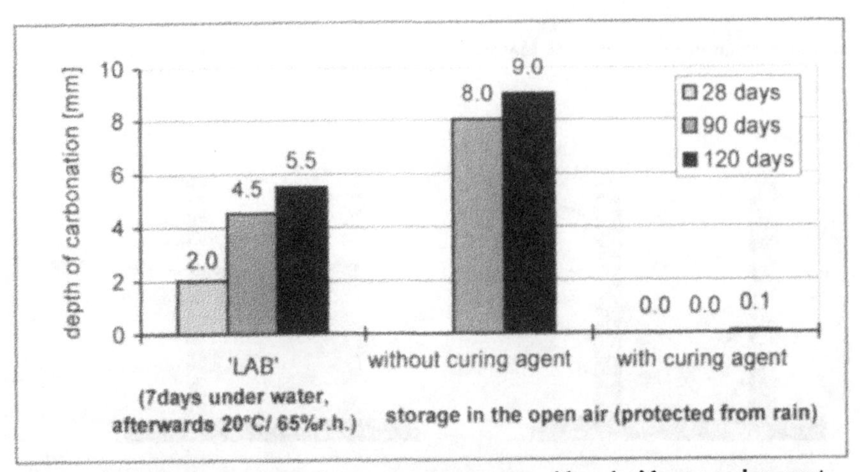

Figure 28. Carbonation of blast furnace cement concrete with and without a curing agent under different storage conditions (cement content = 320 kg/m³, w/c = 0.5) [14]

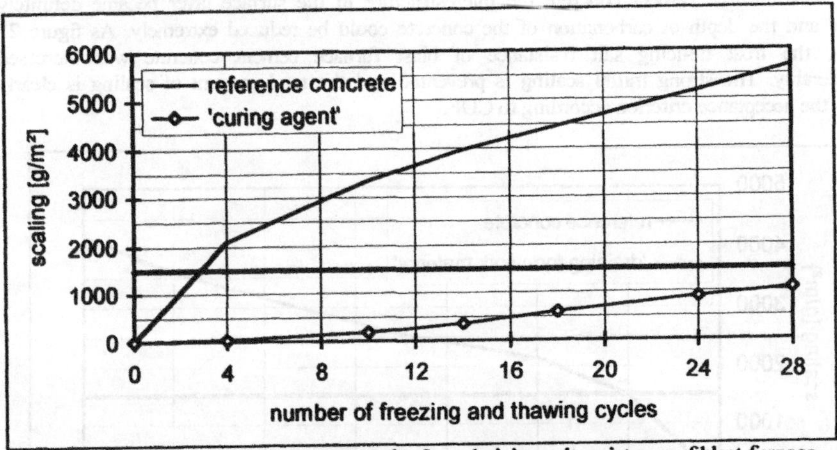

Figure 29. Influence of a curing agent on the frost de-icing salt resistance of blast furnace cement concrete (CDF test [11]; cement content = 350 kg/m³, w/c = 0.5)

REFERENCES

[1] **J. Bijen, "Blast furnace slag cement", BetonPrisma, 's-Hertogenbosch, 1996.**

[2] **F. Cole and B. Kroone, "Carbon dioxide in hydrated portland cement",** ***Journal of the ACI*****, [6], 1275-1294, 1960.**

[3] **J.R. Johnstone and F.P. Glasser, "Carbonation of portlandite single crystals and portlandite in cement paste",** ***Proc. 9th Intern. Congress on the Chemistry of Cement*****, New Dehli, [5] 370-376, 1992.**

[4] S. Matala, "Effects of carbonation on the pore structure of granulated blast furnace slag concrete", PhD thesis, Helsinki University of Technology, Espoo, 1995.

[5] H.M. Ludwig, "Zur Rolle von Phasenwandlungen bei der Frost- und Frost-Tausalz-Belastung von Beton", ("On phase transformation under frost and frost de-icing salt attack on concrete"), PhD thesis, Bauhaus-Universität Weimar, 1996.

[6] H.M. Ludwig and J. Stark, "Freeze-Thaw and Freeze-Deicing Salt Resistance of Concretes Containing Cement Rich in Granulated Blast Furnace Slag", *ACI Materials Journal*, 94 [1] 47-55 (1997).

[7] F. Schröder, "Vaterit, das metastabile Calciumcarbonat, als sekundäres Zementsteinmineral", ("Vaterite, the metastable calcium carbonate, as secondary mineral in cement stone"), *Tonindustriezeitung*, 86 [10] 254-260, 1962.

[8] G. Bayer and H.G. Wiedemann, "Über die Stabilität und das Umwandlungsverhalten des Vaterits ($CaCO_3$)", ("On the stability and transformation of vaterite"); pp. 9-22 in *Angewandte chemische Thermodynamik und Thermoanalytik*, Birkhäuser, Basel, Boston, Stuttgard, 1979.

[9] L. Fernandez-Diaz et.al., "The role of magnesium in the crystallization of calcite and aragonite", *Journal of Sedimentary Research*, 66 [3] 482-492, 1996.

[10] T. Ogino et. al., "The rate and mechanism of polymorphic transformation of calcium carbonate in water", *Journal of Crystal Growth*, 100, 159-167, 1990.

[11] M.J. Setzer, G. Fagerlund and D.J. Janssen, "RILEM Recommendation for Test Method for the Freeze-Thaw Resistance of Concrete - Tests with Sodium Chloride Solution (CDF)", *Materials and Structures* 193 [29] 523-528, 1996.

[12] H.K. Hilsdorf, J. Kropp and M. Günther, "Karbonatisierung und Dauerhaftigkeit von Beton", ("Carbonation and durability of concrete"), pp. 89-96 in *Fortschritte im konstruktiven Ingenieurbau*, Ernst&Sohn, Berlin, 1984.

[13] J. Bonzel and E. Siebel, "Neuere Untersuchungen über den Frost-Tausalz-Widerstand von Beton", ("Last investigation concering the frost de-icing salt resistance of concrete"), *Beton and Stahlbetonbau*, 27 [4] 153-157, 1977.

[14] R. Scholz, "Wirksamkeit von Nachbehandlungsmitteln zur Erhöhung der Dauerhaftigkeit von langsam erhärtenden Betonen", ("Effectiveness of curing agents to improve the durability of slowly hardening concrete"), Master thesis, Bauhaus-University Weimar, 1995.

[15] J. Stark and B. Wicht, "Zement und Kalk. Der Baustoff als Werkstoff.", ("Cement and lime. The building material as material."), pp.376, Birkhäuser, Basel, 2000.

[16] J. Stark and B. Wicht, "Dauerhaftigkeit von Beton. Der Baustoff als Werkstoff.", ("Durability of concrete. The building material as material."), Birkhäuser, Basel, 2001.

EFFECT OF CALCIUM HYDROXIDE CONTENT ON THE FORM, EXTENT, AND SIGNIFICANCE OF CARBONATION

Niels Thaulow, Richard J. Lee, Keith Wagner, and Sadananda Sahu
RJ Lee Group, Inc., Monroeville, PA 15146, USA
(nthaulow@rjlg.com)

ABSTRACT

The literature is incomplete on describing the significance of carbonation as a deleterious phenomenon in concrete, except for its effect on corrosion of reinforcement. The authors present recent observations that indicate that, under certain conditions, carbonation can have a severe negative impact on strength development. The results also indicate that addition of pozzolans to concrete with high water/cement ratio increases the likelihood that carbonation will reduce the strength of concrete.

These results are based on microstructural observations using scanning electron microscopy techniques correlated with microhardness data. It is proposed that, as carbonation proceeds, the reduction in the pH shifts the form of the carbonation from a microscopic precipitation of carbonate (which densifies the paste) to a form which is dominated by growth of discrete calcite crystals in a matrix of silica gel. The strength of the resulting microstructure is extremely dependent on the moisture content of the concrete. This effect is further aggravated by the presence of competing demands for calcium hydroxide, either from pozzolonic activity, leaching, or sulfate attack.

INTRODUCTION

Carbonation of ordinary Portland cement concrete is typically considered to increase the carbonation processes are as follows: strength and decrease the porosity as shown by a number of studies quoted by Parrot [1987, 1990]. The primary disadvantage of the carbonation process is that it lowers the pH of the pore solution. This means that reinforcing steel in carbonated concrete is no longer passivated and active corrosion can take place (Figure 1).

Earlier literature typically described the carbonation process in concrete as a reaction between calcium hydroxide in the cement paste and carbon dioxide from the air [Neville 1996]. It is now well established that the calcium silicate hydrate carbonates, too [St. John et al 1998]. Even unhydrated cement particles can be consumed in the process. The reactions governing the carbonation process are as follows:

(1) $CO_2 + OH^- \leftrightarrows HCO_3^-$
(2) $HCO_3^- + OH^- \leftrightarrows CO_3^{--} + H_2O$
(3) $Ca(OH)_2 + CO_3^{--} \leftrightarrows CaCO_3 + 2OH^-$
(4) $CaO_{1.5}\,SiO_2 \cdot 2H_2O + 1.5CO_3^{--} \leftrightarrows 1.5\,CaCO_3 + SiO_2 + \frac{1}{2}\,H_2O + 3OH^-$

This raises a very important question: "Why would one expect the carbonated concrete to maintain or improve its mechanical and physical properties when all the original binder has been converted or consumed?" The present paper will address this question and show that under certain conditions, the carbonated concrete may actually lose its strength and become as soft as talc.

CARBONATION

Carbonation is a through solution process. Carbon dioxide from the air reacts with hydroxyl ions to form hydrogen carbonate (bicarbonate), HCO_3^-. The bicarbonate ions react with more hydroxyl to form carbonate ions, CO_3^{--}.

At high pH (i.e. high hydroxyl-ion concentration) carbonate ions are the dominant species. They react with calcium ions in solution to precipitate calcium carbonate ($CaCO_3$) which has a low solubility.

The process continues until all available calcium is consumed. The process has a maximum rate at about 60% relative humidity, because the diffusion of carbon dioxide from the ambient air is faster when the capillary pores of the concrete are air-filled. Calcium and hydroxyl ions diffuse to the reaction site through solution. When the ions are consumed in the reaction, solid calcium hydroxide or calcium silicate hydrate goes into solution.

As long as solid calcium hydroxide is available in the system the concentrations of calcium ions and hydroxyl ions remain high. Additionally, calcium carbonate, with low solubility, is precipitated from a supersaturated solution and acts as a good, durable binder in most cases [Thaulow and Jakobsen 1997].

Calcium carbonate is normally found as small calcite crystals in a continuous phase. The crystals nucleate and grow so fast that the silica gel, formed as a byproduct of carbonation of calcium silicate hydrate, is included as impurities in the crystals. Figure 2 shows the microstructure of carbonated concrete as seen in the SEM. Figure 3 shows a sketch of the microstructure of the calcite binder in ordinary carbonated concrete. At a given relative humidity, the rate of carbonation is determined by the water/cement ratio, which controls both the permeability and the relative amount of calcium compounds in the cement

Figure 1. Carbonation-induced corrosion of a bridge pier.

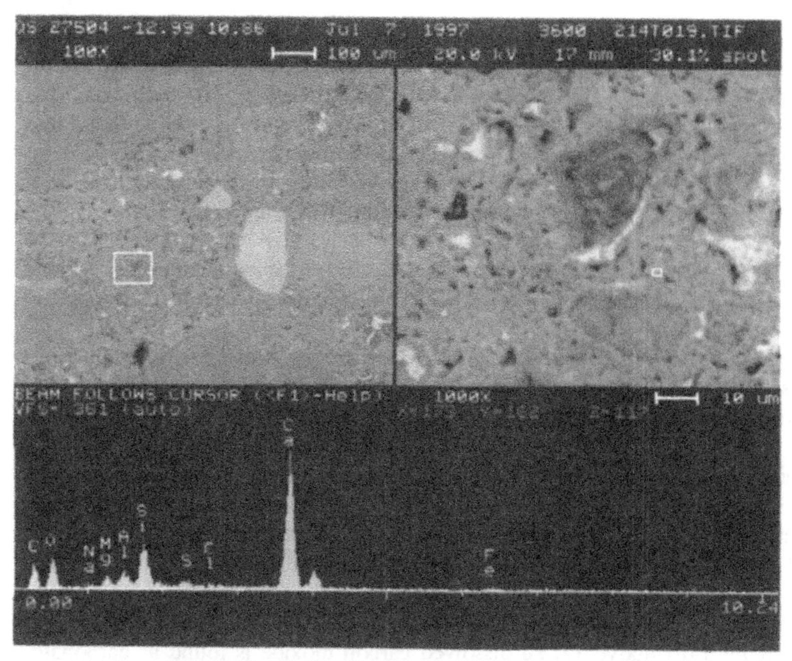

Figure 2 - Carbonated paste

binder. Parrott [1987] quotes results of numerous investigations which have obtained good correlations between carbonation depth and the strength of the concrete. This is basically a water/cement ratio effect. Taylor [1990] notes that carbonation markedly increases the compressive and tensile strengths of Portland cement mortars and concretes, sometimes by as much as 100%, in the affected regions. The strength also increases in slag cement concretes made with as much as 40% slag, but decreases at higher slag contents.

BICARBONATION

In very porous concrete with high water/cement ratio and low content of calcium hydroxide, Thaulow and Jakobsen [1997] found that carbonation could lead to increased porosity and loss of strength. This was first observed in residential concrete basements in Canada, where high water/cement concrete with or without Class C fly ash showed deep carbonation creating a soft, friable layer that could easily be scratched by a knife. In wet conditions, the carbonated zone was even softer and sand grains could be dislodged by a fingernail. Thaulow and Jakobsen [1997] called the process "Bicarbonation" and showed large "popcorn" crystals of calcite in an isotropic matrix of silica gel as seen in the optical microscope.

The present authors have observed the same phenomenon in low quality, high water/cement ratio concrete in residential concrete slabs and foundations in Southern California. The concrete contained Class F fly ash and the water/cementitious ratio was about 0.9 or higher. Figures 4 and 5 show the typical popcorn crystals of calcite formed in this process as seen in the SEM in back-scattered electron mode. The crystals were found as discrete particles in a continuous matrix of silica gel as determined by the EDS analyses.

The resulting effects cause the calcite crystals to act as sand grains and not as binding agents. Ultimately, the new binder then becomes decalcified silica gel.

It is obvious that this resulting microstructure is completely different from what is observed in ordinary carbonation. Figure 6 shows a sketch of the microstructure obtained by this bicarbonation process.

This microstructure and observed crystallite size and morphology indicates that the calcite crystals are growing at low supersaturation instead of high supersaturation. The explanation is found in the speciation of carbon dioxide in aqueous solution depending on the pH as shown in Figure 7 [adopted from Huxley, 2000].

At high pH above 12, all dissolved carbon dioxide is found as carbonate ions, CO_3^{2-}. This is the situation as long as solid calcium hydroxide is present in

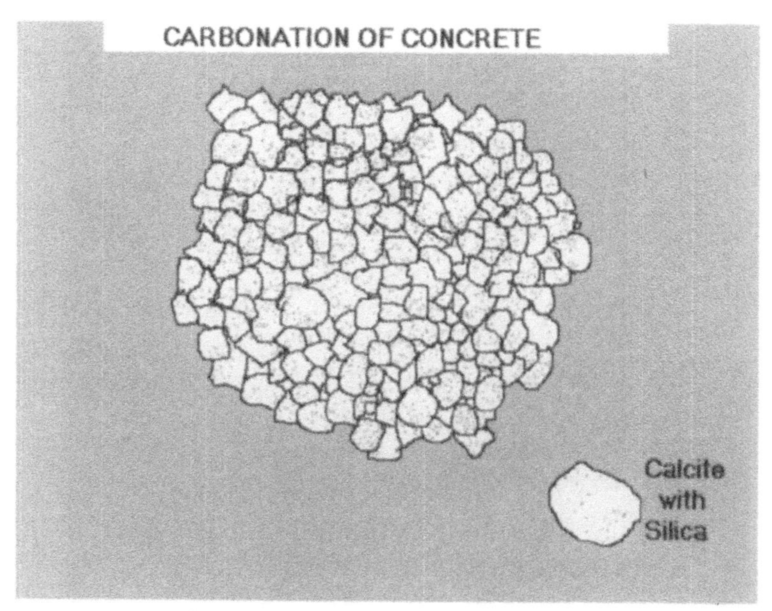

Figure 3. Schematic model of carbonation

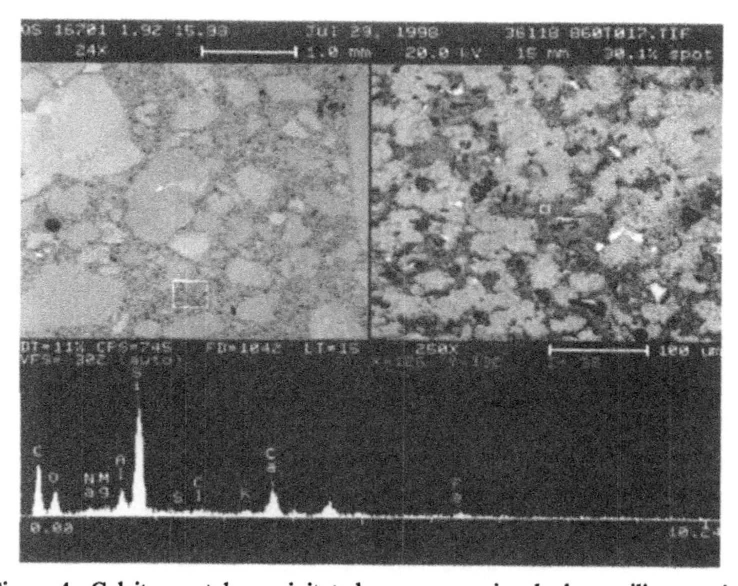

Figure 4. Calcite crystals precipitated as popcorns in a hydrous silica matrix

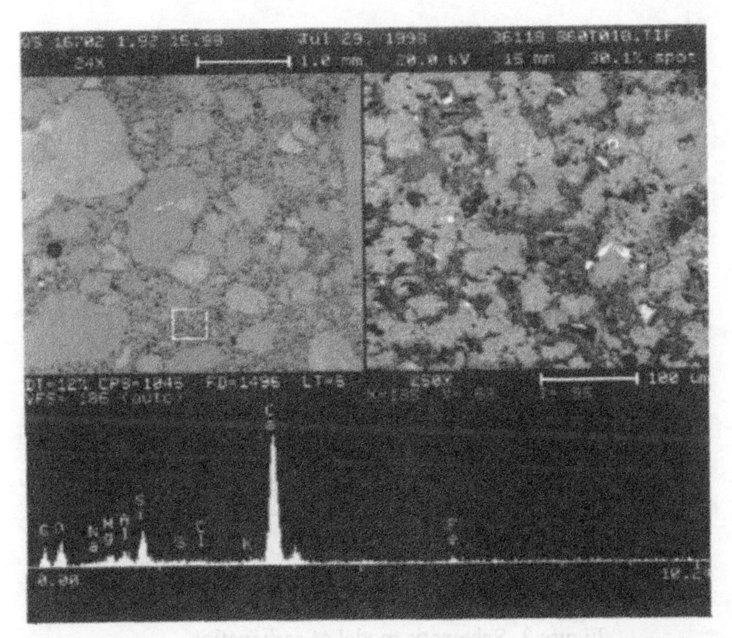

Figure 5. Calcite crystals precipitated as popcorns in a hydrous silica matrix

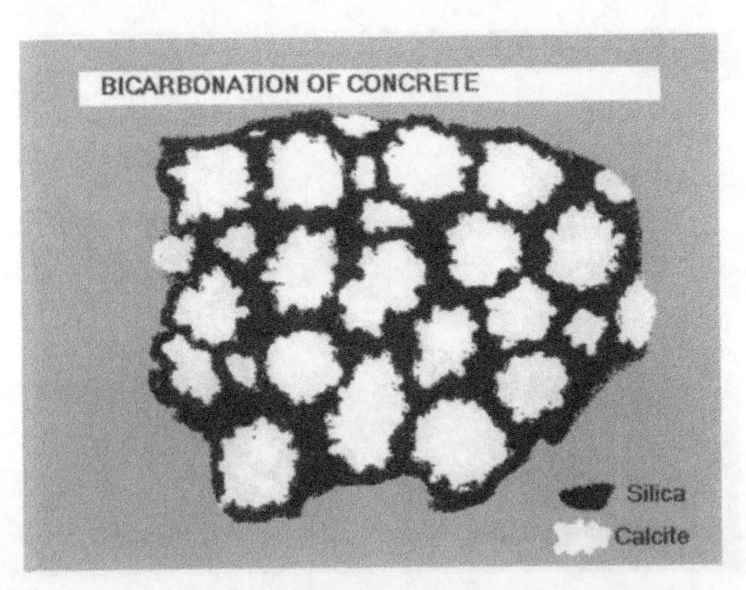

Figure 6. Schematic model of bicarbonation process

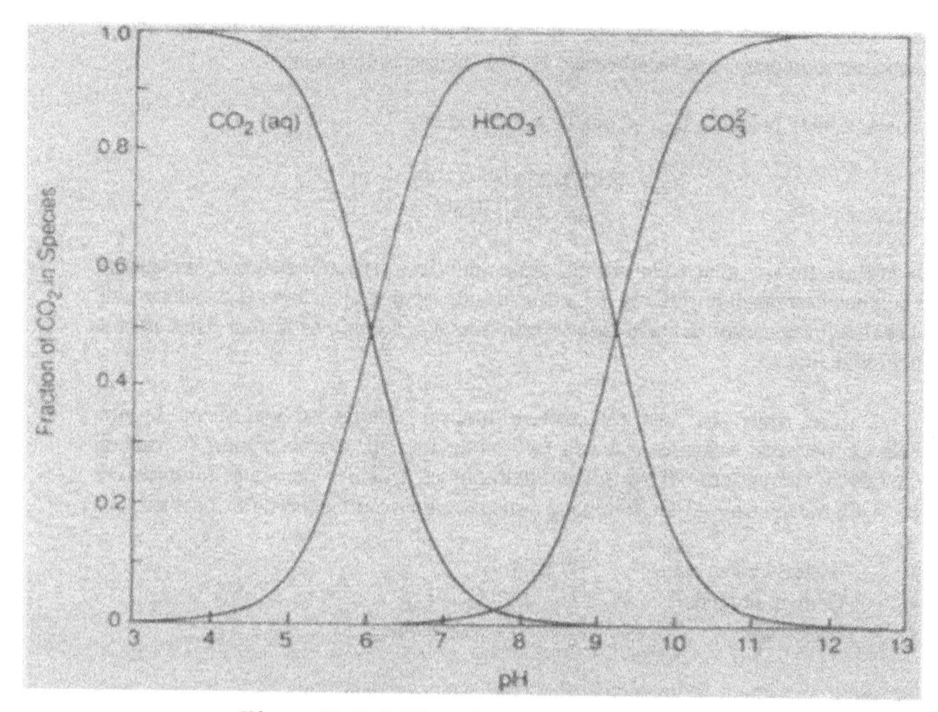

Figure 7. Stability of carbonate species with pH.

the system which maintains the pH at 12.4. At pH below 11 the CO_3^{2-} concentration drops and bicarbonate, HCO_3^-, begins to dominate.

The solubility product, K_{sp}, of calcite is defined as:

$$CaCO_3 \leftrightarrows Ca^{++} + CO_3^{=}$$
$$K_{sp} = [Ca^{++}][CO_3^{=}]$$

It is thus obvious that when the pH drops due to a lack of calcium hydroxide, it becomes increasingly difficult to grow calcite crystals. Ultimately, calcite will dissolve if the carbon dioxide concentration is high enough to further decrease the pH (acid attack).

The route that the carbonation process follows is determined by the balance between available calcium hydroxide and the concentration of carbon dioxide in the system. For a given environment, (relative humidity, temperature and CO_2 concentration) the following concrete parameters govern this balance:

- Water/cement ratio
- Cement content
- Pozzolans

Of these factors it is expected that the water/cementitious ratio is the most important one because it determines both the permeability of the cement paste and the local concentration of calcium hydroxide. The deleterious bicarbonation is not expected to take place at water/cementitious ratio lower that 0.6.

Pozzolans such as fly ash, slag, microsilica, and others may play a double role because they generally decrease the permeability but take away some of the calcium hydroxide reservoir. This may explain why some high slag or high fly ash concrete show poor frost resistance in the carbonated layer.

MICROHARDNESS

The influence of carbonation and bicarbonation on the mechanical properties of the cement paste in concrete exposed in residential concrete has been measured by Vickers microhardness testing. The measurements were taken in both carbonated and non-carbonated concrete under dry and wet conditions. The average of a minimum of ten measurements were taken on each sample per data series.

Figures 8 and 9 show the expected normal increase in strength of carbonated areas over non-carbonated paste for two different sample populations. In both cases the carbonation was of the ordinary carbonation. The dry tests

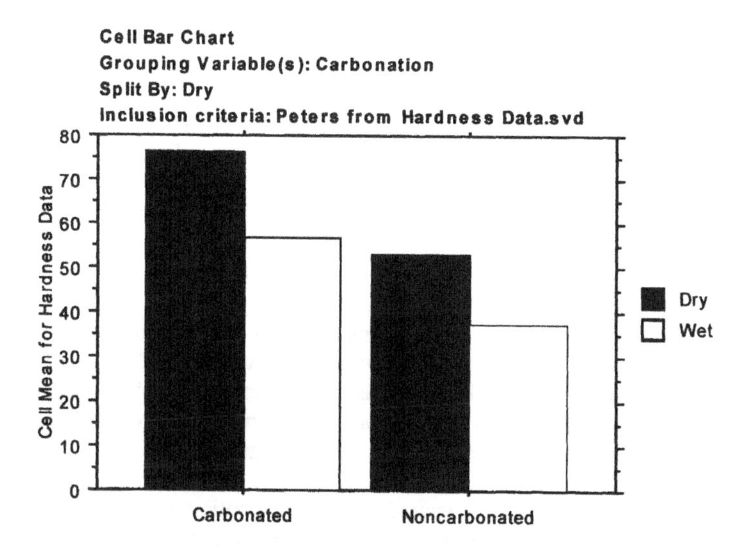

Figure 8. Microhardness data of carbonated concrete.

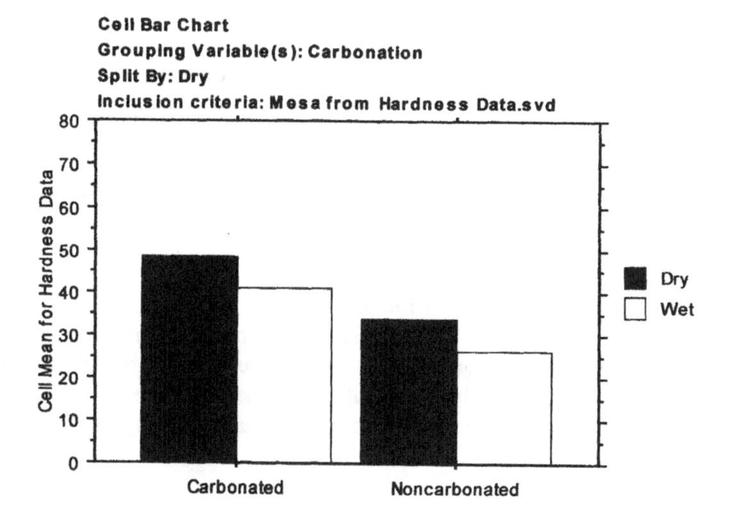

Figure 9. Microhardness data of carbonated concrete.

resulted in higher values for microhardness than the wet tests, both for carbonated and non-carbonated concrete.

Figure 10 shows similar measurements on residential concrete of very high water/cement ratio with fly ash. Water/cementitious ratio was around 0.9 and the fly ash content comprised approximately 20% of the binder. The concrete showed the typical microstructure of bicarbonation with discrete "popcorn" calcite (Figures 4 and 5).

The microhardness of the non-carbonated areas showed low strength and the carbonated areas even lower values. Comparing dry and wet conditions it was found that the non-carbonated paste showed the expected 18% decrease in microhardness. However, in the bicarbonated paste there was a 43% decrease in microhardness between the dry to wet values. The microhardness is as low as the mineral talc, which is the softest on the Mohs hardness scale.

It is believed that this is caused by the imbibation of water into the decalcified silica gel which results in thc extreme softness of the cement paste. This deleterious effect may have been present in earlier studies that showed a similar correlation between strength and carbonation [Parrott, 1987].

CONCLUSIONS

Field observations on low quality, high water/cement ratio residential concretes sometimes show unusually soft and friable carbonated surfaces. The microstructures of the carbonated layers show bicarbonation resulting in scattered calcium carbonate crystals in decalcified silica gel. This process results in a severe loss of mechanical properties as documented by microhardness measurements.

The deleterious bicarbonation process occurs when the availability of calcium hydroxide in the concrete is too low to maintain the pore liquid at pH sufficiently high to have high concentration of carbonate ions compared to bicarbonate ions during carbonation. The process may occur at high water/cement ratios and may be aggravated by the presence of competing demands for calcium hydroxide either from pozzolanic activity, leaching, or sulfate attack.

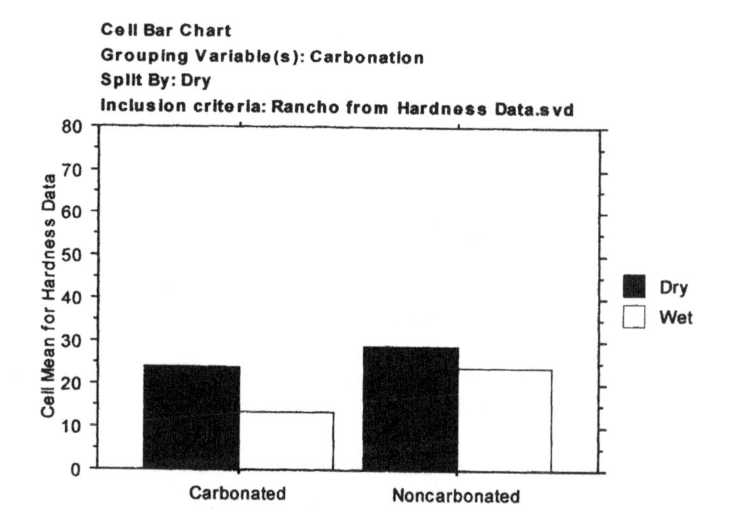

Figure 10. Microhardness data of bicarbonated concrete.

REFERENCES

Huxley,2000
http://www.Huxley.ic.ac.uk/Local/EarthSciUg/ESFirstyr/EarthMaterials,
Imperial College, London

Neville, A.M., 1996, "Properties of Concrete"
John Wiley & Sons, London, 496-506

Parrott, L.J., 1987, "A Review of Carbonation in Reinforced Concrete"
UK, Cement and Concrete Association, 42

Parrott, L.J., 1990, "Damage Caused by Carbonation of Reinforced Concrete"
Materials and Structures 23 (135), 230-234

St. John D.A., Poole A.W., and Sims I, 1998, "Concrete Petrography"
A handbook of investigative techniques, Arnold, London, 206

Taylor, H.F.W., 1990, "Cement Chemistry"
Academic Press, London, 386

Thaulow, N. and Jakobsen U.H., 1997, "Deterioration of Concrete Diagnosed by Optical Microscopy", Proceedings of the Sixth Euroseminar on Microscopy Applied to Building, Materials. Reykjavik, Iceland, 282-296.

SIMPLISTIC TESTS DO NOT NECESSARILY REVEAL THE COMPLETE PICTURE OF THE CARBONATION OF $Ca(OH)_2$ IN FIBRE-CEMENT PRODUCTS

J W Figg[1], W J French[2], A M Hutchinson[3] and M B Willoughby[3]

(1) John Figg & Associates, 5 Andrewes Croft, Great Linford, Milton Keynes MK14 5HP, UK. (2) Geomaterials Research Services, Falcon Park, Crompton Close, Basildon SS14 3AL, UK. (3) Eternit UK Ltd, Meldreth, nr Royston SG8 5RL, UK.

ABSTRACT

In a case involving fibre-cement sheet roofing of an insulated store, evaluation of carbonation via determination of reduction in alkalinity (phenolphthalein spray) and change in birefringence (polarizing microscopy) gave poor agreement. It is suggested that crystallisation of high-birefringence calcium formate by reaction of calcium hydroxide with formaldehyde/formic acid derived from the insulation is the probable cause of the anomalous results.

INTRODUCTION

This study was commenced to obtain a better understanding of apparently anomalous carbonation behavior of fibre-cement corrugated roofing sheets which had been installed on the roof of a steel-framed cool store building at Holbeach, Norfolk for a little under ten years [1,2]. The 150mm corrugation sheets had been laid at 15 degrees pitch and the exposed (outside) surfaces had been left as manufactured. The interior (underside) of the roof sheets had been given one coat of paint (specified to be chlorinated rubber) and then a 100mm thickness of polyurethane foam insulation had been applied by spraying.

fibre-cement sheets, nor does it correspond with current advice on the use of coatings with thin cementitious materials [3, 4].

This is because where thin sheet materials are concerned it is important that both sides of the sheets shall be treated equivalently or the different conditions of exposure of the two sides can lead to distortion and cracking caused by unequal moisture movement or shrinkage through differential carbonation of the cement paste.

THE SITE SITUATION

The ridge of the cool store building at Holbeach was oriented approximately east-west i.e. there was a south slope and a north slope to the roof. The roof of the main part of the building was coated on the underside with chlorinated rubber paint and insulated with the spray-applied polyurethane foam. An extension to the slope of the roof on the north side formed a lean-to open-ended storage area. The fibre-cement sheets of the lean-to roof were left as manufactured on both sides.

Roof sheets on the insulated part of the building had developed cracks parallel to the corrugations and leakage of rainwater into the building had been reported. No such cracking had been observed in the sheets of the uninsulated lean-to roof.

For more than 30 years it has been known that single-sided sealing of, particularly new, fibre-cement (asbestos-cement) sheets and the resultant differential carbonation, can lead to warping and eventually cracking of the material because of the unequal shrinkage stresses that are created. Damage is also usually exacerbated by differential moisture and thermal movement stresses.

Thus it was anticipated that examination of the corrugated roofing sheets would confirm this preliminary diagnosis.

THE SAMPLES AND THEIR EXAMINATION

100 mm diameter disc samples were cut from the roof using a diamond barrel core drill. The samples were taken in accordance with the following scheme:

from the south slope of the roof
from the north slope of the roof
from the northern lean-to roof
within this general plan particular conditions were selected:
cracked sheets

uncracked sheets
making sure that the locations included:
samples from the east end of the roof
samples from the west end of the roof
and also including samples of:
side laps between sheets
end laps of sheets
the central portion of sheets
duplicate specimens were taken for each sample condition.

A total of 72 disc (core) samples were drilled.

Control samples of new fibre-cement sheet were also obtained from current production.

TESTS ON SITE SAMPLES

The disc samples of fibre-cement sheet were broken in half and one of the freshly-broken surfaces was sprayed with an aqueous ethanolic solution of phenolphthalein [5,6,7]. The magenta-coloured portion of the material was taken to represent uncarbonated cement paste. The mean thicknesses of the uncoloured (carbonated) zones at the top and bottom surfaces of the sheet were estimated to the nearest 0.1 mm.

Table 1 Result of carbonation measurements using phenolphthalein indicator (mm)

Sample Type	South Slope	North Slope	Lean-to Roof
Exposed surface	1.1	1.3	1.4
Underside	0.3	0	1.2

(Mean of 12 samples for N and S slopes, mean of 4 samples for lean-to roof)

Another piece from selected samples (one sample for each exposure condition) was used to prepare a thin-section (0.02 mm thickness) for petrographic examination. The thin-sections were examined using a polarizing microscope with the polars partially-crossed. The thicknesses of the bright birefringent zones at the top and bottom surfaces of the fibre-cement sheet was determined using a calibrated graticule. These zones were assumed to be formed by carbonation of the cement paste [7].

Table 2 Results of carbonation measurements using polarized light microscopy (mm)

Sample Type	South Slope	North Slope	Lean-to Roof
Exposed surface	0.8	1.1	0.9
Underside	1.0	1.0	2.0

(Mean of 3 samples for S slope, mean of 4 samples for N slope, single sample for lean-to)

The paint coating on the underside of the roof, although specified to be a chlorinated rubber, was found to be brittle rather than flexible and no chlorine could be detected by Beilstein's test. It appeared that an alternative paint system had been applied.

EVALUTION OF RESULTS OF TESTS ON SITE SAMPLES

With the main part of the building, where the underside of the roof has been coated with the paint/polyurethane foam combination, carbonation has been prevented, whereas the exposed surface has carbonated (somewhat less on the drier south slope than on the wetter and cooler north slope).

The phenolphthalein test results for the lean-to samples show that, in this instance, carbonation is more or less equal on the underside and the exposed surface. Normally carbonation of the undersurface would be expected to be greater than for the top (exposed) surface since the rate of carbonation is greatest at Relative Humidities of about 50% whilst the exposed surface can be expected to be frequently either much wetter or much drier than this RH. Probably this is a sampling effect.

The optical microscopy test results (Table 2) are anomalous in that the coated undersurfaces of the insulated roof appear to show significant carbonation, to more or less the same extent as the exposed upper surface of the roof. On the other hand the microscope measurements do give a carbonation depth for the underside of the lean-to roofing greater than for the upper surface. The unexpected lack of correlation between the two techniques required further consideration.

In reality neither test method actually measures carbonation, rather the effects produced by carbonation. In the case of the phenolphthalein method it is the reduction in alkalinity that is revealed. With polarized light microscopy it is the pronounced change in birefringence that is characterized as being indicative of the formation of calcium carbonate.

The phenolphthalein test results are typical of the findings for Portland cement-based materials exposed to the normal atmosphere (which contains approximately 0.035% carbon dioxide). But the results of the optical microscopy measurements need explanation.

Calcium carbonate (calcite) has one of the largest birefringence values ($n_\varepsilon - n_\omega$ = 0.172) for natural materials [8], whereas hydrated cement paste has a low birefringence (of the order of 0.005 - 0.008). Consequently uncarbonated cement paste appears largely dark when viewed with crossed polars in a petrographic microscope, whilst carbonated cement paste appears bright. However, there are other possible chemical reactions that could result in large differences in birefringence.

Polyurethane foam tends to evolve formaldehyde (this effect has been blamed as one cause of "sick-building" syndrome). It has also been alleged that formaldehyde generation followed by oxidation to formic acid has been the cause of deterioration of museum metallic artifacts (especially lead alloys) that have been displayed or stored in cabinets constructed of polyurethane-bonded plywood or chipboard.

Although the paint used on the soffit of the corrugated roof sheets was not positively identified, it is well documented that the formaldehyde molecule is small and can penetrate relatively thin organic coatings. Furthermore, calcium formate has the relatively high birefringence value of 0.068.

It was postulated, therefore, that the anomalous results of the optical microscopy measurements might be the result of calcium formate crystallization rather than carbonation. Laboratory tests were consequently commenced to evaluate this hypothesis.

TESTS ON FIBRE CEMENT SAMPLES EXPOSED TO ACIDIC VAPOURS

100 x 80 mm samples of fibre-cement flat sheet (made with both Portland cement and Portland cement/pulverised fuel ash binders) were exposed above acidic solutions contained in glass jars fitted with rubber gaskets and a clamp arrangement to hold the sheet samples at a distance of 25 mm above the liquid surface. The test solutions used were:

- water
- 40% formaldehyde solution (formalin)
- 40% formic acid solution
- 40% acetic acid solution

Water was included as a control and acetic acid because calcium acetate has a cubic structure and hence should not exhibit birefringence.

The relatively concentrated solutions were selected in order to accelerate the test program to fit in with the on-going litigation timetable. Exposure times of 1 and 3 weeks and 3 months were adopted to compare with unexposed samples of fibre-cement material.

RESULTS OF EXPOSURE TESTS

Formic acid and to an even greater extent, acetic acid, vapours caused blistering of the fibre-cement material, an effect that was obvious after only one week's exposure and was dramatic at the end of the test at 3 months.

The blistering was caused by the formation of lenses of crystalline material. Thin sections of the test specimens revealed that these lenticular bulges were birefringent, the formic acid samples showing the brightest colours, but even the acetic acid samples revealed some birefringence. Clearly the chemical reactions are more complex than simple salt formation by calcium hydroxide with the ingressing vapours.

The accelerated tests resulted in a coarse-structure alteration product morphology that differed considerably from the normal appearance of carbonated cement paste. In any case carbonation involves shrinkage of cement paste not expansion as produced by the organic acids.

However, the samples exposed over formalin solution were not blistered but a birefringent layer developed which had similarities with carbonated fibre-cement sheet material. The alteration layer in samples where pulverised fuel ash was included in the material was twice the depth compared with the sheet materials made without PFA.

DISCUSSION

The laboratory tests show that volatile acidic vapours can produce high-birefringence zones in cementitious materials, especially fibre-cement sheets, that are not dissimilar to the appearance of carbonated cement paste.

It is possible, therefore, that the unusually thick bright zones detected during thin-section polarized light examination of the site samples could be due to the effects of formaldehyde/formic acid migration from the spray-applied foam insulation.

Of course, the experimental variables did not include the presence of a paint coating (chlorinated rubber or other) on the fibre-cement surfaces exposed to the solutions in the test jars (to include this would have doubled the number of samples and, in any case, would probably have required an unacceptably long exposure time).

Also, since no accelerated carbonation samples were examined in these tests, it is not possible to comment on the appearance of such carbonated cement paste under the polarizing microscope. It is likely that any speeded-up reaction will result in a coarser microstructure of the alteration product.

For litigation purposes it may be sufficient to throw doubt on the routine technique used for comparing the microscopic and phenolphthalein methods for evaluating carbonation, but further work is needed to confirm that the seeming carbonation zone is due to calcium formate rather than calcium carbonate (or, of course, a mixture of both).

X-Ray Diffraction analysis should be able to distinguish possibilities, but to date it has not been feasible to undertake this research and so this paper takes the opportunity to draw attention to a practical problem that clouded the issues in an already complex legal case and to remind researchers and investigators that neither pH evaluation nor birefringence are the whole story in the reaction of alkaline cementitious materials with acidic vapours.

What the overall investigation did confirm, however, is that unequal exposure of thin materials can lead to formation of alteration products that can result in distortion, tensile stress development and cracking. Whatever fibrous reinforcement is include in cement composites it is important to remember that it is the cement matrix that changes dimensions in response to environmental influences, the fibres just reinforce.

Perhaps the final irony is that none of this information was eventually used in the litigation case!

REFERENCES

[1] Designer's Duty of Skill and Care: Implied Terms of Fitness for Purpose and Merchantable Quality; *QV Limited and another v Frederick F Smith & others,* Construction Industry Law Letter, July/August 1998, 1998 CILL 1397-1408, pp1403-1405.

[2] Rudi Klein, Contractual Matters: Some Lessons on Design Responsibility, M&E Design, December 1998, pp 10-11.

[3] Painting Asbestos Cement, Ministry of Public Building and Works, Advisory Leaflet 28, second edition 1967.

[4] Painting Walls, Part 1: Choice of Paint, Building Research Establishment, Digest 197, 1982.

[5] M H Roberts, Carbonation of Concrete Made with Dense Natural Aggregates, Building Research Establishment, Information Paper IP 6/81, April 1981, 4 pp.

[6] Carbonation of Concrete and its Effects on Durability, Building Research Establishment, Digest 405, May 1995,8 pp.

[7] I Sims, The Assessment of Concrete for Carbonation, Concrete, Nov/Dec 1994, Vol 28, No 6, pp 33-38.

[8] Michel-Lévy Color Chart, Carl Zeiss Ltd, Oberkochen, Germany.

ETTRINGITE – FRIEND OR FOE?

H.F.W. Taylor

Maundry Bank, Lake Road, Coniston, Cumbria, LA21 8EW, UK

ABSTRACT

Ettringite is in some cases associated with expansion and cracking in cementitious materials, but in others it is a major hydration product which contributes to the formation of a strong and durable material. Factors governing the effects of ettringite formation on strength and expansion are discussed, including the roles of calcium hydroxide and alkali.

INTRODUCTION

It is sometimes thought that the large increase in solid volume that accompanies ettringite formation is always likely to cause expansion, of which it is a sufficient explanation. However, the increase when ettringite is formed from C_3A and gypsum is similar to that which occurs on hydration of C_3S (Fig. 1). Both processes are, in the broad sense, hydration reactions, in which water is transferred from the liquid into one or more solid products. The solid volume clearly increases, while the total volume decreases because of the tighter packing of the hydrogen and oxygen atoms in the products. In a non-porous material, an increase in solid volume necessarily implies one in total volume, but in a porous material, such as

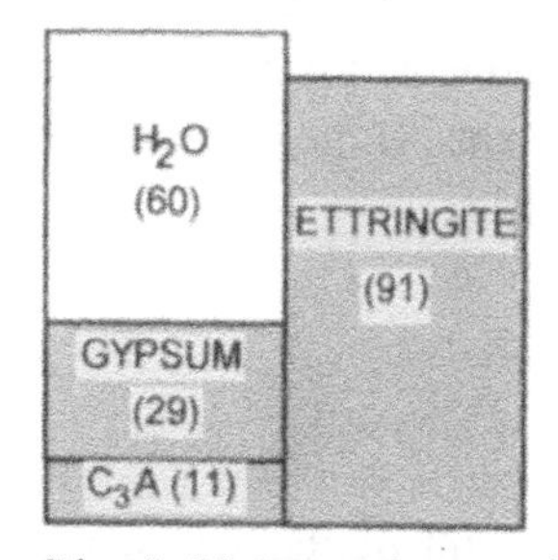

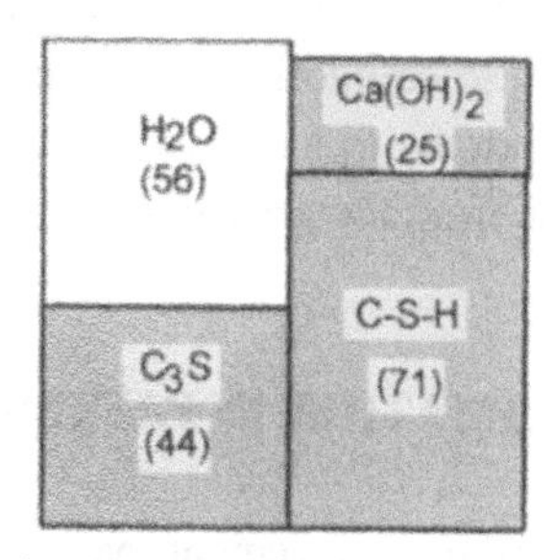

Fig. 1. Volume changes in ettringite formation and in C_3S hydration.

cement paste, some or all of the increase can be accommodated in the pores. More positively, without an increase in solid volume, there could be no development of strength.

Increases in strength and expansion are not mutually exclusive. In a saturated paste of portland cement, strength development is accompanied by a small volume increase. Both the increase in strength and the expansion that result from an increase in solid volume are determined by such factors as the microstructure of the material, where in the microstructure the product is formed, how much of the product is formed, the stresses that its formation produces in the surrounding material and the manner in which that material responds to those stresses. If the resistance to the stresses is sufficient, the reaction stops. This situation, which is one in which strength increase is dominant, occurs in the hydration of a cement paste of low water/cement ratio. If the resistance is not sufficient to stop the reaction, the strength increases until the available space has been occupied, after which damaging expansion can take place.

CRYSTAL GROWTH

Although other mechanisms are possible,[1] we shall assume that pressures caused by ettringite formation result from crystal growth. The driving force for growth from solution is supersaturation, and any growth pressure produced cannot exceed a value set by the degree of supersaturation. A crystal will not grow where it has to exert pressure if it can do so in another direction where this is unnecessary, or if another crystal can grow in its stead without exerting pressure. High growth pressures are therefore most likely to occur in small, confined spaces where there is also a high degree of supersaturation. Scherer[2] has shown that, if certain equilibrium conditions are satisfied, the growth of a crystal in a pore can generate a substantial pressure only if the pore radius is well below 100 nm. It is uncertain whether these equilibrium conditions would exist for an ettringite crystal growing in close proximity to a source of aluminate ions. If they do not, substantial pressures might arise from the growth of larger crystals.

ETTRINGITE FORMATION IN PASTES

Ettringite formation in a cement paste that is still plastic will not result in significant expansion. This could largely account for the fact that substantial quantities of ettringite can be formed during the early hydration of portland cement without producing expansion. It is also unlikely to produce significant expansion if it is formed while there is a sufficient quantity of free space, and this condition may well persist after the paste has developed considerable strength. Under these conditions, ettringite formation may contribute to development of strength. However, if the same amount of ettringite is formed in a mature paste, in which there is little free space, it may well cause expansion and accompanying decrease in strength.

Although the changes in solid and total volumes associated with ettringite formation are similar to those occurring on hydration of C_3S, formation of ettringite in a given material does not have the same effect as that of the same volume of C_3S hydration products. If more than a certain amount of ettringite is formed in a portland cement paste, damaging expansion occurs, but the formation of C–S–H and calcium hydroxide continues to produce strength with only minor expansion until no more space is available and reaction stops. Evidently, in the case of C_3S hydration, the stresses produced are smaller, or the ability of the material to resist those stresses is greater, or both.

SOME CASES IN WHICH ETTRINGITE FORMATION CONTRIBUTES TO STRENGTH WITHOUT PRODUCING MAJOR EXPANSION

Supersulfated Cements

These are mixtures of granulated blastfurnace slag (80–87%), anhydrite (10–15%) and portland cement or clinker (<5%). Interest in them has waned during the past 30 years, and data on their hydration are incomplete, but several aspects appear well established.[3–5] By 3 days, substantially all the anhydrite has reacted and much ettringite and some C–S–H have formed. Subsequently, more C–S–H forms, while the content of ettringite seems to remain substantially unchanged. Some, at least, of the ettringite takes the form of needles, which may form a three-dimensional meshwork. TEM replicas showed needles up to 120 μm long[6], but results obtained by this technique may not be representative.

Table I gives the assumptions made in a very approximate calculation of the hydration stoichiometry at an age of 3 days and the results obtained. For a water/cement ratio of 0.5, and assuming a density of 2000 m^3kg^{-1} for the C–S–H gel, a slightly more elaborate calculation taking into account the formation of a hydrotalcite-type phase and the reaction of the portland cement clinker gives the volume composition indicated in Fig. 2a.

At later ages, more C–S–H gel forms. Its Ca/Si ratio will still be low, but may be higher than that of the material formed earlier because ettringite formation is no longer contributing to the demand for CaO. It is uncertain what happens to the Al_2O_3 that is released; probably, a substantial fraction of it enters the C–S–H gel. No significant quantity of calcium hydroxide is formed at any stage.

Opinions have differed as to the relative importance of the ettringite and C–S–H gel in relation to strength development. It is reasonable to suppose that this is largely attributable to ettringite initially, but that the C–S–H becomes increasingly important and ultimately dominant.

Table I. Calculated stoichiometry for a 3-day old paste of a supersulfated cement.

Assumptions

The cement contains 85% of slag (41% CaO, 33% SiO_2, 12% Al_2O_3, 8% MgO), 12% of anhydrite and 3% of portland cement clinker.

The anhydrite reacts completely to form ettringite.

The amount of slag reacting is that needed to supply the Al_2O_3 for the ettringite.

The SiO_2 from the slag, and the CaO not required to form ettringite, react to form C–S–H gel with a H_2O/SiO_2 ratio of 4.

Conclusions

The principal constituents, in g/100g of cement, are ettringite, 37 g; C–S–H gel, 23 g; unreacted slag, 60 g.

The C–S–H gel has a Ca/Si ratio of ~0.7.

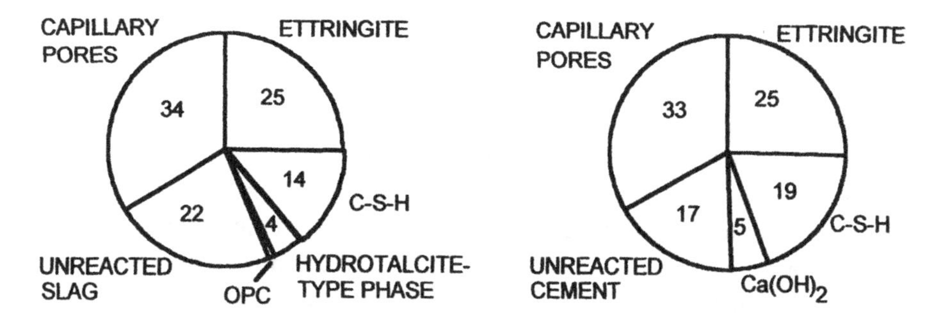

A. Supersulfated cement B. Oversulfated portland cement (7.1% SO_3)

Fig. 2. Calculated volume fractions in two 3-day old pastes (w/c ratio 0.5).

Sulfoaluminate-Belite Cements

These contain $C_4A_3\overline{S}$ and belite as major constituents, often with other phases, such as ferrite. Like supersulfated cements, they form substantial quantities of ettringite initially, which is increasingly accompanied by C–S–H.[7,8] In the case of a Chinese cement studied by Wang Lan and Glasser[8], the ettringite soon decreased in amount, and was almost completely replaced by AFm phases by 28 days. Calcium hydroxide was also present at early ages, but it, too, was resorbed by 28 days. As with supersulfated cements, the ettringite present at early ages forms an interlocking network composed of needles up to a few micrometers in length, which is probably largely responsible for the binding.

Blending with lime and calcium sulfate can render sulfoaluminate-belite cements expansive. Sudoh et al.[7] observed that the ettringite crystals formed at early ages did not then form a network, but were concentrated around the clinker grains, and considered that this was a probable cause of expansive behavior.

MIXTURES OF CALCIUM ALUMINATE CEMENTS WITH CALCIUM SULFATE

Bayoux et al.[9] studied the hydration of a calcium aluminate cement in mixtures with anhydrite, hemihydrate or gypsum and in ternary blends including calcium hydroxide. The most clear cut results related to pastes made with calcium aluminate cement and anhydrite. Additions of 9% or 19% of anhydrite produced strength increases with no significant expansion, and XRD showed the formation of ettringite, CAH_{10} and C_4AH_{13}. A 29% addition led to expansion and cracking, and XRD showed ettringite and AH_3 as the only detectable hydration products. In no case was calcium hydroxide detected. Limited replacement of calcium aluminate hydrates by ettringite in the products thus increased strength, but complete replacement caused expansion.

CEMENTS FOR SUPPORTING ROADWAY WALLS IN COAL MINES

These are based on the use of two slurries, which are separately pumped to the place in which the cement is required, at which point they are mixed.[10] One slurry contains a source of aluminate ions ($C_4A_3\overline{S}$ or a calcium aluminate cement), and the other contains lime and calcium sulfate. The slurries also contain various admixtures and bentonite is added to one to keep the solids in suspension while the slurry is being pumped. The slurries are reasonably stable until they are mixed, when they rapidly produce ettringite.

The products are very open meshworks of ettringite needles, individually some 5 µm long. The compressive strengths are only around 5 MPa, but this is sufficient for the purpose for which the cements are designed. The observations confirm that ettringite crystals, formed where adequate free space is available, can form meshworks that have measurable strength. The meshworks appear to resemble those formed at early ages from supersulfated and sulfoaluminate-belite cements, but do not have an infill of other phases. The products contain little or no calcium hydroxide.

SOME CASES IN WHICH ETTRINGITE FORMATION CAUSES SIGNIFICANT EXPANSION

Oversulfated Portland Cements

It is well known that excessive contents of total SO_3 in a portland cement cause expansion, but the literature contains little detailed information on the chemistry and almost nothing on the microstructural changes that occur during hydration of such cements. Fig. 2b shows the results of a tentative calculation of the phase distribution by volume in a paste of water/cement ratio 0.5, in which the amount of SO_3 was chosen to produce the same amount of ettringite as in the supersulfated cement paste discussed earlier (Fig. 2a). The two distributions are remarkably similar, though the content of hydration products other than ettringite is somewhat higher for the oversulfated portland cement.

The oversulfated portland cement paste would probably expand, whereas that of the supersulfated cement does not. One can only speculate on the reason for the difference, which is clearly not due to one in the contents of ettringite at 3 days. Faster formation of C–S–H in the case of the oversulfated portland cement may decrease the space available to the ettringite. A three dimensional network of ettringite crystals is probably formed in the case of the supersulfated cement but not in that oversulfated portland cement. Calcium hydroxide is present only in the case of the oversulfated portland cement. The degree of supersaturation could be higher in the case of the oversulfated portland cement.

External Sulfate Attack; Delayed Ettringite Formation (DEF) in Heat Cured Materials

An XRD study confirmed the widely accepted view that ettringite is formed when portland cement concrete is attacked by Na_2SO_4 solutions, and showed that its formation is accompanied by consumption of calcium hydroxide.[11] An SEM investigation led to similar conclusions, but the only evidence for the presence of the ettringite was provided by microanalyses that were compatible with mixtures on a sub micrometer scale of ettringite with C–S–H and in some cases also with gypsum.[12] Similar evidence of ettringite formation was obtained for a paste attacked by $MgSO_4$ solution,[12] but in this case the major cause of damage is destruction of C–S–H gel. In another SEM study of $MgSO_4$ attack, only minor quantities of ettringite were detected.[13]

In the case of DEF, there have been differences of opinion as to whether the ettringite crystals that cause expansion are ones of sub micrometer size in the paste or much larger crystals formed in cracks at aggregate interfaces and elsewhere.[14] The former view is the more probable, the larger crystals being in this case recrystallization products. The very small crystals in the paste are probably formed in close admixture with C–S–H and monosulfate in the outer product, in confined and poorly connected spaces at high levels of supersaturation. This situation would be closely similar to that which appears to exist in attack by Na_2SO_4 solutions. In both cases, calcium hydroxide is present, at least when ettringite formation begins. The quantities of ettringite present in materials that have expanded after heat treatment are often higher than those in identical materials that have been cured at ordinary temperature.

Mixtures of $C_4A_3\overline{S}$, Calcium Sulfate and Calcium Hydroxide

These serve as a model for an important class of expansive cements. The constituents are the same as those in some of the cements designed for use in coal mines mentioned earlier, but we shall consider here their behavior when used in pastes at ordinary water/cement ratios. If the proportions are suitably chosen, ettringite is the sole product:

$$C_4A_3\overline{S} \quad + \quad 8C\overline{S}H_2 \quad + \quad 6CH \quad + \quad 74H_2O \quad \Rightarrow \quad 3C_6A\overline{S}_3H_{32} \qquad (1)$$

At normal water/cement ratios, an unrestrained paste of this composition expands markedly. Initially, very small, unoriented ettringite crystals are formed at the surfaces of the $C_4A_3\overline{S}$ particles (Fig. 3). Subsequently, radial aggregates of prismatic ettringite crystals appear, and typically grow to a length of some 50 μm. Expansion occurs when these aggregates impinge.[15,16] The exact seat of expansion is uncertain, but is likely to be where the supersaturation is highest. Possibilities include the inner ends of the needles, and the very small, unoriented crystals of ettringite, which might continue to form as precursors.

If the mixture is cured in a completely sealed mould, so as to prevent expansion, strengths similar to those of comparable portland cement pastes are obtained.[17] The acicular morphology of the ettringite crystals is poorly developed and there appears to be a greater proportion of randomly oriented crystals between the radial aggregates. If a calcium aluminosilicate glass is substituted for the

$C_4A_3\bar{S}$, the paste does not expand significantly, even if it is unrestrained, and the ettringite crystals are unoriented.[17]

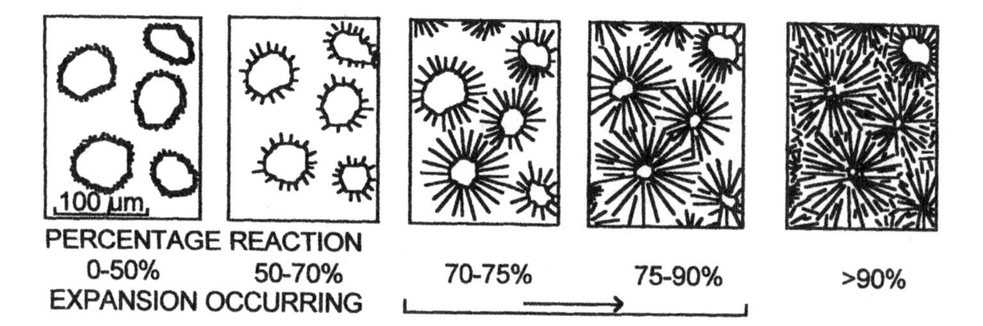

Fig. 3. Growth of ettringite crystals around $C_4A_3\bar{S}$ (after Ogawa and Roy[16]).

THE ACTION OF CALCIUM HYDROXIDE

The results that have been described show that expansion associated with ettringite formation tends strongly to be associated with the presence of calcium hydroxide, though this is not an absolute requirement. Mehta[18] noted that, in the presence of calcium hydroxide, ettringite tends to form as stubby crystals approximately 1 µm long, whereas in its absence it forms longer needles. He suggested that the small crystals imbibed water and that this was the cause of expansion. However, it is not obvious why ettringite crystals, even if considerably smaller than those mentioned above, should attract water molecules more strongly than does C–S–H, which has a much larger specific surface area. Of course, the formation of ettringite entails uptake of water molecules, irrespective of the mechanism by which it occurs, but this is not the same thing as attraction of water to crystals that have already been formed.

Okushima et al.[19] and Nakamura et al.[20] developed a different hypothesis. The precipitation of ettringite from solution may be represented by the equation

$$6Ca^{2+} + 2Al(OH)_4^- + 4OH^- + 3SO_4^{2-} + 26H_2O \Rightarrow Ca_6[(Al(OH)_6]_2(SO_4)_3 \cdot 26H_2O \quad (2)$$

The concentration of $Al(OH)_4^-$ in solution is very low, and if those of the other ions are relatively high the $Al(OH)_4^-$ ions will not migrate far from the solid phase providing them before they react to form ettringite. In the presence of calcium hydroxide, these conditions are met, and the ettringite forms in regions of high supersaturation close to the aluminate source, so that growth pressures can be high. In the absence of calcium hydroxide, the $Al(OH)_4^-$ ions migrate more freely. The ettringite crystals are dispersed throughout the paste, where the degree of supersaturation is lower, and growth pressures are low.

On this hypothesis, the morphology of the ettringite crystals does not determine the expansive behavior but is a consequence of the conditions under which they are formed. High supersaturation would be expected to yield a large number of small crystals and low supersaturation fewer and larger crystals.

EFFECTS OF ALKALI HYDROXIDES

The hypothesis outlined above does not take account of the fact that the pore solutions of portland cement pastes more than a few days old are essentially ones of alkali hydroxides, with only low concentrations of Ca^{2+}. This might be considered to invalidate it. However, some observations on DEF in heat treated mortars suggest a resolution of the problem.

Famy[21] and Famy et al.[22] found that ettringite formation and expansion are relatively rapid if storage at ordinary temperature after the heat treatment is in water, markedly slower if it is in a simulated pore solution and insignificant at 200 days if it is in a more concentrated KOH solution. During storage in water, leaching occurs, and much alkali hydroxide is lost from the paste. Ettringite formation and expansion are thus restricted until the pH has been reduced by leaching. Loss of OH^- ions also occurs in external sulfate attack, at least under some conditions.[12]

Equation (3) represents approximately the metastable equilibrium that exists between ettringite and other species in an alkaline environment:

$$Ca_6[(Al(OH)_6]_2(SO_4)_3{\cdot}26H_2O \quad + \quad 4OH^- \quad \Leftrightarrow$$
$$Ca_4[(Al(OH)_6]_2(SO_4){\cdot}6H_2O \; + \; 2Ca(OH)_2 \; + \; 2SO_4^{2-} \; + \; 20H_2O \quad (3)$$

Alkali drives the equilibrium to the right, inhibiting the formation of ettringite.[23]

Deng and Tang[24] found that addition of 1–3% of Na_2O to a supersulfated cement led to strong expansion and that the addition of 0.12% of K_2O to a sulfoaluminate cement paste sigificantly increased expansion. They considered that the increased OH^- concentration restricted the migration of $Al(OH)_4^-$ away from its source. It is difficult to reconcile this hypothesis with observations that the $Al(OH)_4^-$ concentration is markedly increased in the presence of alkali, and further investigation is required. The observations could possibly be explained by effects on the rates of reaction of the anhydrous phases.

CONCLUSIONS

- Ettringite can both contribute to strength development and produce expansion. No single factor determines the relative importance of these effects in a given case; rather, several factors are involved.
- Strength development from ettringite formation is favoured by weakly reactive sources of aluminate ions, by absence of calcium hydroxide and by a sufficiency of free space when the ettringite is formed.
- Under these conditions, the degree of supersaturation is low and the ettringite crystals are widely dispersed throughout the paste. They are typically needles several micrometers long and in some cases, at least, they form a three dimensional network.
- Expansion from ettringite formation is favoured by highly reactive sources of aluminate ions, by the presence of calcium hydroxide and by restriction of space when the ettringite is formed.
- Under these conditions, the degree of supersaturation is high and the ettringite crystals are formed close to the aluminate source. In some and possibly all cases, the ettringite crystals that cause the expansion are of sub micrometer dimensions.
- Alkali hydroxides present after heat treatment of portland cement materials retard and restrict expansion from delayed ettringite formation, but are reported to increase expansion in normally cured pastes of supersulfated and sulfoaluminate cements.

ACKNOWLEDGMENT

I thank Dr K.L. Scrivener for helpful discussions.

REFERENCES

[1]P.W. Brown and H.F.W. Taylor, "The Role of Ettringite in External Sulfate Attack"; pp. 73–97 in *Materials Science of Concrete: Sulfate Attack Mechanisms*, Edited by J. Marchand and J.P. Skalny. American Ceramic Society, Westerville, Ohio, 1999.

[2]G.W. Scherer, "Crystallization in Pores," *Cement and Concrete Research,* **29** [8] 1347–1358 (1999).

[3]R. Kondo and S. Ohsawa, "Studies on a Method to Determine the Amount of Granulated Blastfurnace Slag and the Rate of Hydration in Slag Cements"; pp. 255–262 in Vol. 4 of *Proceedings of the 5th International Symposium on the Chemistry of Cement*. Cement Association of Japan, Tokyo, 1969.

[4]J.C. Yang, "Chemistry of Slag-Rich Cements"; pp. 296–309 in Vol. 4 of *Proceedings of the 5th International Symposium on the Chemistry of Cement*. Cement Association of Japan, Tokyo, 1969.

[5]K. Mohan and S.P. Ghosh, "Evaluation of Physio-Chemical and Hydration Characteristics of Supersulfated Cement"; pp. 331–337 in Vol. 3 of *9th International Congress on the Chemistry of Cement, New Delhi, India, 1992*. National Council for Cement and Building Materials, New Delhi, 1992.

[6]H.G. Midgley and K. Pettifer, "The Micro Structure of Super Sulphated Cement," *Cement and Concrete Research,* **1** [1] 101–104 (1971).

[7]G. Sudoh, T. Ohta and H. Harada, "High-Strength Cement in the $CaO-Al_2O_3-SiO_2-SO_3$ System and its Application"; pp. V-52–V-157 in Vol. 3 of *7th International Congress on the Chemistry of Cement*. Editions Septima, Paris, 1980.

[8]Wang Lan and F.P. Glasser, "Hydration of Calcium Sulphoaluminate Cements," *Advances in Cement Research,* **8** [31] 127–134 (1996).

[9]J.P. Bayoux, A. Bonin, S. Marcdargent and M. Verschaeve, "Study of the Hydration Properties of Aluminous Cement and Calcium Sulphate Mixes"; pp. 320–334 in *Calcium Aluminate Cements*, Edited by R.J. Mangabhai. E. and F.N. Spon, London, 1990.

[10]S.A. Brooks and J.H. Sharp, "Ettringite-Based Cements"; pp. 335–349 in *Calcium Aluminate Cements,* Edited by R.J. Mangabhai. E. and F.N. Spon, London, 1990.

[11]J.G. Wang, "Sulfate Attack on Hardened Cement Paste," *Cement and Concrete Research*, **24** [4] 735–742 (1994).

[12]R.S. Gollop and H.F.W. Taylor, "Microstructural and Microanalytical Studies of Sulfate Attack. I. Ordinary Portland Cement Paste," *Cement and Concrete Research*, **22** [6] 1027–1038 (1992).

[13]D. Bonen and M.D. Cohen, "Magnesium Sulfate Attack on Portland Cement Paste. I. Microstructural Analysis," *Cement and Concrete Research*, **22** [1] 169–180 (1992).

[14]C. Famy, K.L. Scrivener and H.F.W. Taylor, "Delayed Ettringite Formation;" in *Structure and Performance of Cements*, 2nd ed. Edited by J. Bensted and P. Barnes. Spon, London, in press.

[15]A. Bentur and M. Ish-Shalom, "Properties of Type K Expansive Cement of Pure Components. II. Proposed Mechanism of Ettringite Formation and Expansion in Unrestrained Paste of Pure Expansive Component," *Cement and Concrete Research*, **4** [5] 709–721 (1974).

[16]K. Ogawa and D.M. Roy, "$C_4A_3\overline{S}$ Hydration, Ettringite Formation, and Its Expansion Mechanism. II. Microstructural Observation on Expansion. **12** [1] 101–109 (1982).

[17]I. Odler and P. Yan, "Investigations on Ettringite Cements," *Advances in Cement Research*, **6** [24] 165–171 (1994).

[18]P.K. Mehta, "Mechanism of Expansion Associated with Ettringite Formation," *Cement and Concrete Research*, **3** [1] 1–6 (1973).

[19]M. Okushima, R. Kondo, H. Muguruma and Y. Ono, "Development of Expansive Cement with Calcium Sulphoaluminous Cement Clinker"; pp. 419–438 in Vol. 4 of *Proceedings of the 5th International Symposium on the Chemistry of Cement*. Cement Association of Japan, Tokyo, 1969.

[20]T. Nakamura, G. Sudoh and S. Akaiwa, "Mineralogical Composition of Expansive Cement Clinker Rich in SiO_2 and Its Expansibility"; pp. 351–365 in Vol. 4 of *Proceedings of the 5th International Symposium on the Chemistry of Cement*. Cement Association of Japan, Tokyo, 1969.

[21]C. Famy, "Expansion of Heat-Cured Mortars"; Ph.D. Thesis, Imperial College, University of London, 1999.

[22]C. Famy, K.L. Scrivener, A. Atkinson and A.R. Brough, "Influence of the Storage Conditions on the Dimensional Changes of Heat-Cured Mortars," *Cement and Concrete Research*, in press.

[23]W. Wieker and R. Herr, "Zu einigen Problemen der Chemie des Portlandzements" [On Some Problems of the Chemistry of Portland Cement], *Zeitschrift für Chemie*, 29 [9] 321–327 (1989).

[24]Deng Min and Tang Mingshu, "Formation and Expansion of Ettringite Crystals," *Cement and Concrete Research*, 24 [1] 119–126 (1994).

Workshop participants in discussion

THE ROLE OF CALCIUM HYDROXIDE IN ALKALI RECYCLING IN CONCRETE

Michael Thomas
University of Toronto

ABSTRACT

It is generally accepted that $Ca(OH)_2$ plays an important part in alkali-silica reactions (ASR) in concrete although the precise role has not been clearly defined. The author of this paper has previously demonstrated that ASR is relatively benign if calcium is not readily available in a mortar or concrete. In this paper it is shown that calcium freely exchanges with alkalis in the original alkali-silica reaction product thereby recycling the alkalis and making them available for further reaction with thermodynamically unstable silica in the aggregate. "Alkali"-silica gel found in samples taken from a 55-year-old concrete dam was found to contain considerably lower contents of alkali than that found in younger laboratory samples, the overall composition of the mature gel being very close to that of the C-S-H found in the same concrete. The phenomenon of alkali recycling may explain why continued expansion is often observed in old structures even when laboratory tests (e.g., pore solution analysis, soluble alkali content determination, etc.) indicate that there is little potential for further expansion to occur (due to the low availability of alkali in the system). This is also consistent with the observation that the expansion of concrete in laboratory tests may continue long after the composition of the pore solution has reached apparent equilibrium. In other words the attainment of a low and steady alkali concentration in the pore solution (typically between 0.20 to 0.30 Moles per litre) of concrete containing reactive aggregate does not necessarily mean that the reaction has come to a standstill, but that the rate of consumption of alkali by the reacting silica is in balance with the rate of alkali release from the alkali-silica reaction product (due to exchange with calcium). It is conceivable that in large structures, ASR is not limited by the availability of the alkalis (beyond the initial "activation" of ASR). In such cases the availability of reactive silica or "soluble" calcium may be the limiting factor.

INTRODUCTION

The role played by calcium hydroxide in the deleterious expansion of concrete due to alkali-silica reaction has been debated by a number of authors in the open literature and the following hypotheses have been put forward:

- Calcium may replace alkalis in the reaction product thereby regenerating alkalis for further reaction (Hansen, 1944)
- $Ca(OH)_2$ may act as a buffer maintaining a high level of OH^- in solution (Wang and Gillott, 1991)
- High calcium concentrations in the pore solution prevent the diffusion of silica away from reacting aggregate particles (Chatterji, 1979; Chatterji and Clausson-Kass, 1984)
- If calcium is not available reactive silica may merely dissolve in alkali hydroxide solution without causing damage (Diamond, 1989)
- The formation of calcium-rich gels is necessary to cause expansion either directly or through the formation of a semi-permeable membrane around reactive aggregate particles (Thomas et al. 1991; Bleszynski and Thomas, 1998; Thomas, 1998).

In this paper the role played by calcium in the phenomenon of "alkali recycling", first postulated more than 50 years ago (Hansen, 1944), is further examined using data from laboratory and field studies of ASR-affected concrete. An attempt is made to use alkali recycling to explain observations of continued long-term expansion in field concretes in the apparent absence of "available alkalis" determined on the basis of alkalis in water extracts or in the concrete pore solution.

BRE STUDIES ON 7-YEAR-OLD CONCRETE PRISMS

Approximately 10 years ago the writer and co-workers reported results from a study on a series of concrete prisms that had been stored for seven years on an exposure site at the Building Research Establishment (BRE) in the U.K.; details of the study have been reported elsewhere (Thomas et al, 1991). The prisms were produced using concrete containing a reactive flint sand (from the Thames Valley), different contents of high-alkali Portland cement and varying levels of fly ash (from bituminous coal). The study included an examination of thin sections by petrographic microscopy and polished samples by scanning electron microscopy.

Microprobe (wavelength dispersive x-ray) analysis was performed on some of the polished sections using a Cambridge Instruments Microscan 9.

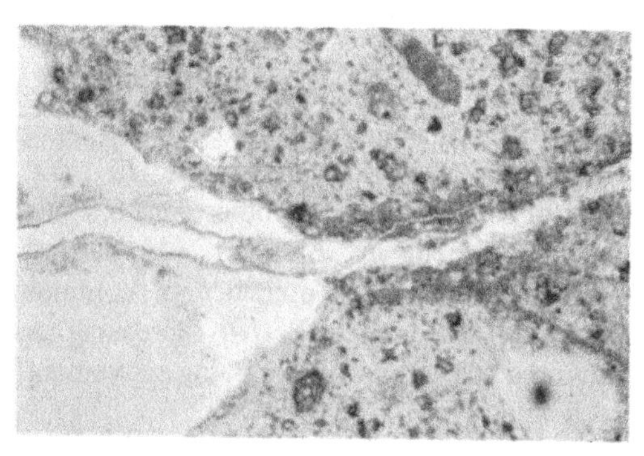

Figure 1 Thin section showing darkening of paste adjacent to gel-filled crack

Figure 1 is a micrograph from a thin section showing a crack running through a reactive flint particle into the adjacent paste. The cement paste immediately adjacent to the gel-filled crack exhibits a darkened zone consistent with the depletion of calcium hydroxide. This is a common observation with ASR-affected concrete. It can also be observed from this micrograph that the visual appearance of the gel changes as the crack extends from the flint particle into the cement paste. The change in visual appearance was accompanied by a change in the elemental composition as determined by microprobe analysis. A summary of the data for the gel found in Portland cement concretes is given in Table 1 and individual data points are shown in Figure 6. Full details of the analyses are given in a previous paper (Thomas et al, 1991).

Table 1 Composition of Gel found in 7-Year-Old Portland Cement Concrete Containing Flint Sand

Location of gel	No. of analyses	Atomic Ratio	
		Ca/Si	K/Si
Within flint boundaries	23	0.25	0.20
In cement paste	35	0.92	0.10

It is clear that the gel in the paste is higher in calcium and lower in potassium compared with the gel found within the boundaries of the flint particle. If it is assumed that the gel in the flint particle represents an earlier stage in the evolution of the reaction product than these data indicate that calcium has exchanged for potassium in the gel. Sodium contents were not reported as part of this earlier analysis as this element was at the limit of the detection capabilities of the microprobe and there was a low confidence in the accuracy of the results.

OBSERVATIONS FROM ASR-AFFECTED DAMS

Data will be presented here from studies of two ASR-affected hydraulic dams both of which contain greywacke-argillite as the reactive aggregate. One of these dams (DS) is located in the U.K. and was approximately 25 years old at the time of sampling and the other dam (CN) is in Canada and was about 55 years old when the field investigation was carried out. At the time of sampling both structures, there was evidence of continuing expansion in the form of (i) crack widening (DS and CN), (ii) increasing operation problems such as sluice gate jamming (CN), and (iii) ongoing movement of the structure indicated by surveying techniques (DS).

Figures 2 and 3 show the results of expansion tests carried out on cores cut from the structure. In both cases, little expansion was observed in the standard test condition (over water at 38°C), but there was significant expansion if a supply of alkali was provided by immersing cores in 1 M NaOH at the same temperature.

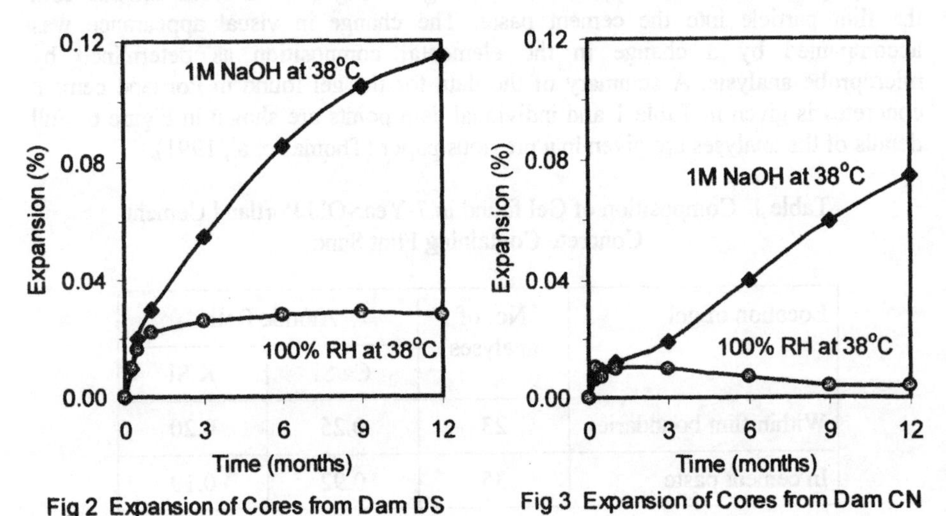

Fig 2 Expansion of Cores from Dam DS

Fig 3 Expansion of Cores from Dam CN

Significant expansion of cores when immersed in alkali solution indicates that a supply of reactive silica remains in the concrete, however, the absence of expansion for cores stored over water suggests that there is an insufficient quantity of alkali available (in the concrete) for further reaction. Further evidence of the low alkali availability was provided by chemical analysis of the hardened concrete and of pore solution extracted from the concrete. Details regarding the analyses and the procedures used for resaturating the cores are provided elsewhere (Thomas, 1996). Data are presented in Table 2.

Table 2 Alkali Contents in Hardened Concrete and Pore Solution of Concrete taken from Hydraulic Structures

Dam	Na_2O_e (% mass)		Core No.	Pore solution (mM/litre)		
	Water-soluble	Acid-soluble		Na^+	K^+	OH^-
DS	0.07	0.12	1	49	71	130
			2	60	82	138
CN	0.08	0.10	-	-	-	-

The chemical analysis reveals that between 1.6 to 1.9 kg/m^3 of alkali (as Na_2O_e) is in a water-soluble form in these concretes (measured concrete densities were in the range of 2350 to 2400 kg/m^3). The pore solution data indicate that only 0.4 to 0.6 kg/m^3 of alkali are actually in the pore solution (measurement indicated water contents in the range of 90 to 130 litres of evaporable water per m^3 concrete).

Diamond has postulated that a hydroxyl ion concentration of around 0.25 Moles/litre (~ pH 13.4) is necessary to sustain the alkali-silica reaction in concrete (Diamond et al, 1981; Diamond, 1983). Kolleck and co-workers (1986) suggested a threshold concentration of 0.20 Moles/litre (~ pH 13.3). In proposing a method for determining the residual reactivity in ASR-affected concrete structures, Thaulow and Geiker (1992) suggested that pore solution measurements provide the best indication of the remaining alkali availability, and claimed that alkali in solution below 0.2 Moles/litre (hydroxide) is not available for reaction.

On the basis of these previous studies, it can be concluded that there is insufficient alkali available in the concrete from the two dams studied to promote further reaction. This is contrary to field observations at the time of sampling (i.e. there was evidence of continued concrete expansion).

Figure 4 below shows a back-scattered electron image (BSI) of a polished sample prepared from a concrete core taken from dam DS. The image shows a gel-filled crack running from a reactive aggregate particle (top right) into the cement paste. The gel within the aggregate appears to have different structure to the

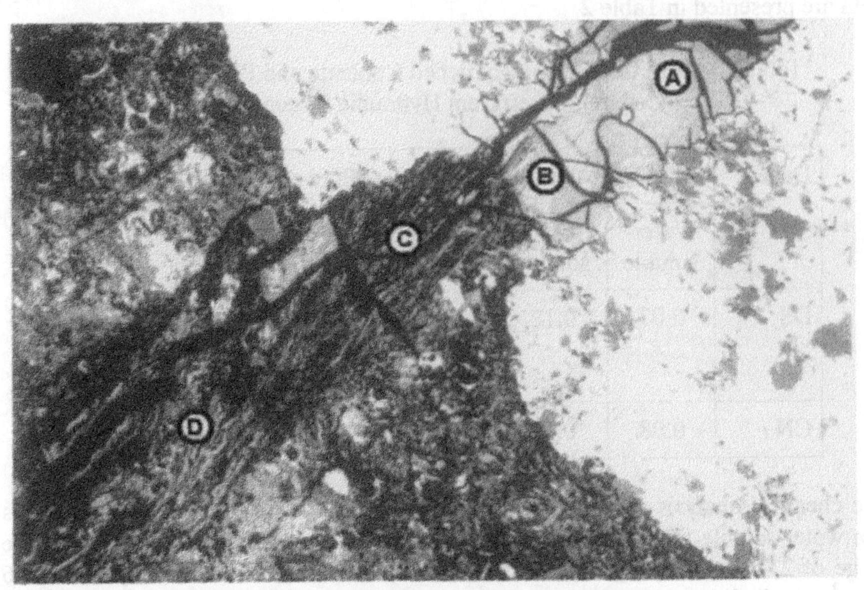

Figure 4 BSI of Polished Sample from Dam DS

product that has formed within the crack in the paste. Table 3 presents the results of energy-dispersive x-ray analysis for four points labeled A to D in Figure 4.

Table 3 Results of EDX Analysis for Points A to D in Figure 4

Location (Fig. 4)	Atomic Ratio	
	Ca/Si	(Na + K)/Si
A	0.21	0.23
B	0.29	0.13
C	1.24	0.09
D	1.41	0.06

These data indicate a definite compositional difference in the gel found in different locations and further supports the hypothesis that calcium from the cement paste substitutes for alkali in the gel.

Figure 5 shows a backscattered electron image and x-ray dot maps for a polished sample from dam CN. The BSI shows a greywacke particle occupying the lower left-hand side of the image and two cracks emanating from the particle into the cement paste (top right-hand side of image). The dot map at top right (Si) in

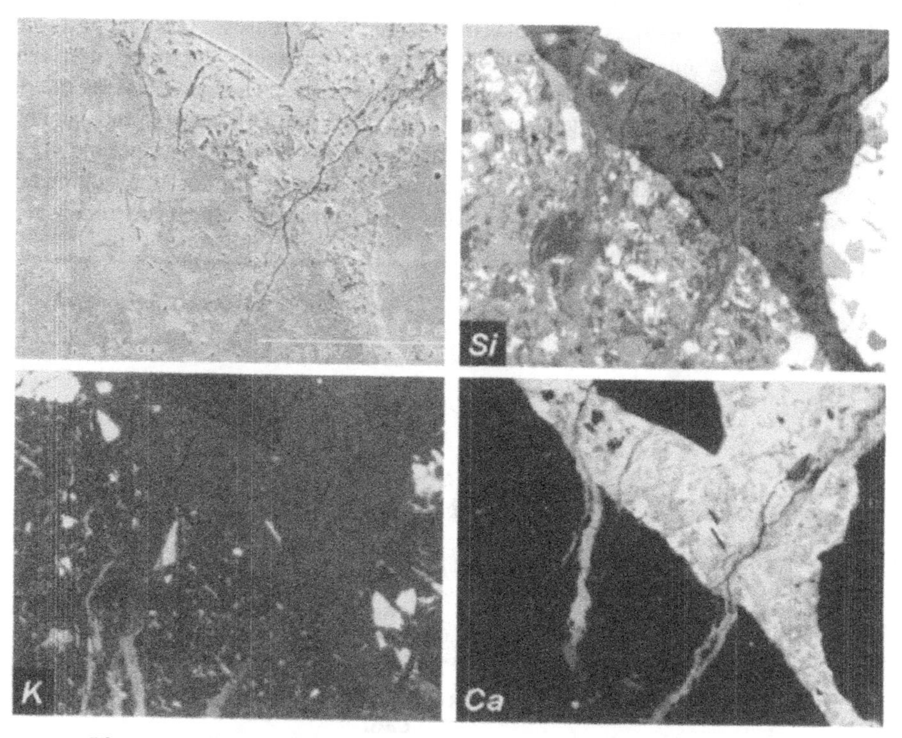

Figure 5 Back-Scattered Image (top left) and X-Ray Dot Maps for Polished Sample from Dam CN

Figure 5, shows little contrast in the silicon content of the product in the crack as it moves from the aggregate into the paste. The dot maps at bottom left (K) and bottom right (Ca) indicate that the gel in the crack changes in composition from a potassium rich to a calcium rich product. This transformation does not occur at the aggregate-cement interface but within the aggregate about 150 microns from the interface. This indicates that the calcium from the paste has slowly penetrated into the gel-filled crack in the aggregate, replacing the potassium (and sodium) as it goes.

Figure 6 shows a comparison of the composition (atomic ratios K/Si vs. Ca/Si) for gels found in the 7-year-old concretes discussed above and the 55-year-old concrete dam (CN). Also shown are some data for the C-S-H "inner product" in the concrete dam. The composition covers a wide range, but it appears that there is a reasonable relationship between the potassium and calcium contents, i.e. as the calcium content decreases the potassium content decreases. This supports the concept of a cation exchange with the calcium replacing the potassium (and sodium) in the initially alkali-rich reaction product. This exchange occurs as the gel migrates away from the aggregate particle and comes into contact with the calcium-rich cement paste. The data in Figure 6 indicate that the process continues slowly as the concrete ages and that the composition of the reaction product may ultimately approach something similar to C-S-H. This "final product" appears to retain very

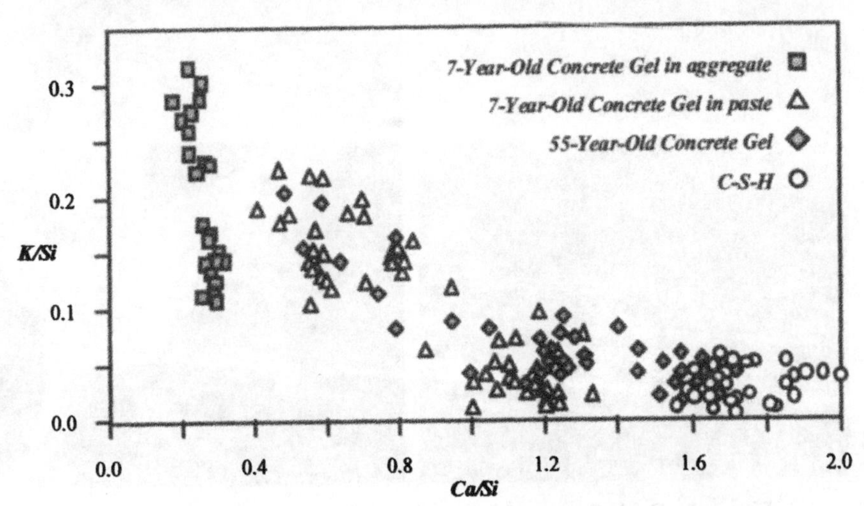

Figure 6 Results of Micro-Analysis for 7-Year-Old Laboratory Concretes and 55-Year-Old Concrete from Dam CN

little alkali indicating that most of alkali that originally participated in the reaction process may eventually be released possibly to participate in further reaction.

RELATIONSHIP BETWEEN PORE SOLUTION & EXPANSION IN CONCRETE CONTAINING REACTIVE AGGREGATE

This section presents data from expansion and pore solution studies carried out at BRE by the author and co-workers; details of the laboratory work and some of the expansion data have been reported previously (Thomas et al. 1996). Figure 7 shows data for concrete specimens cast with reactive flint sand (as 25% of the total mass aggregate) and high alkali cement (1.15% Na_2O_e). Pore solution studies were conducted on concrete specimens containing 400 kg/m^3 of cement and a water-cement ratio of W/C = 0.50. Cubes (100 mm) were stored sealed in plastic at 20°C until extraction and analysis of the pore solution at different ages. The data show that the pore solution alkalinity drops steadily to an age of 1 year after which appears to remain stable with a hydroxyl ion concentration of 0.26 to 0.27 Moles/litre. Also shown is the expansion of concrete prisms, which were wrapped in moist toweling and plastic, and were stored at 20°C. It is clear that the expansion of concrete continues long beyond the point at which the alkali concentration in the pore solution reaches a steady concentration.

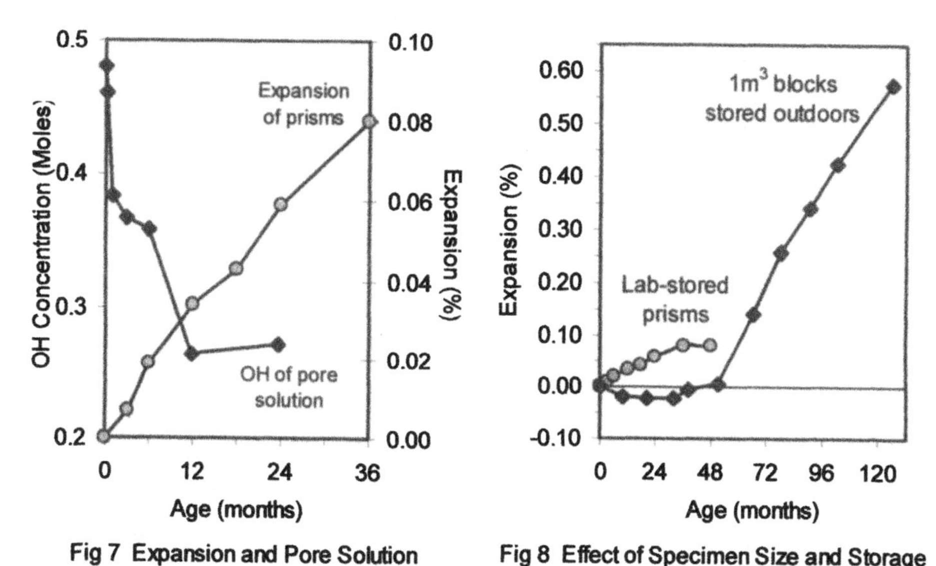

Fig 7 Expansion and Pore Solution Results for Concrete with Flint Sand

Fig 8 Effect of Specimen Size and Storage on Expansion of Concrete with Flint Sand

Figure 8 compares the expansion of concrete prisms stored in the laboratory at 20°C with the expansion of large concrete blocks (~ 1 m^3) stored outdoors at BRE;

both concretes contain 475 kg/m^3 of high-alkali cement. The smaller specimens expand rapidly during the first few years' storage under laboratory conditions, but reach a maximum expansion of around 0.08% after just 3 years. The larger specimens shrink slightly during the first 4 years and then show a steady and continuing expansion for the next 6 years. At an age of 10.5 years the blocks exhibit an average expansion of close to 0.6% and appear to be continuing to expand.

DISCUSSION

The data presented here indicate that the concentration of alkalis in the concrete pore solution is not necessarily a good indicator of the potential for further reaction. Although the pore solution may eventually attain a steady alkali concentration this does not mean that the alkali-silica reaction has ceased, but that there is a balance between the alkalis consumed by new reaction and those released from the reaction product. The fate of the alkalis in concrete containing a source of reactive silica may be explained by reference to Figure 9.

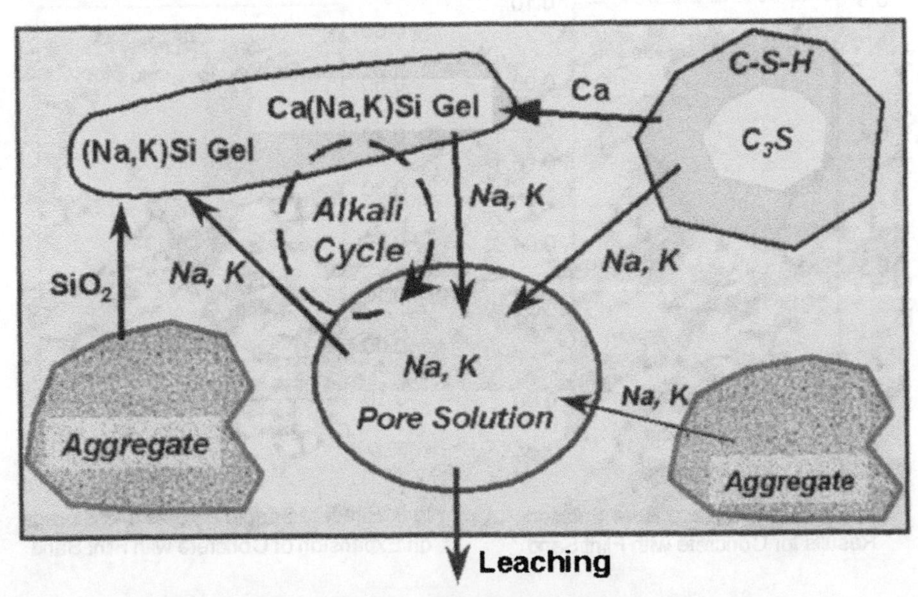

Figure 9 Schematic showing the Fate of Alkalis in Concrete Containing Reactive Aggregate

The primary source of the alkalis in the concrete pore solution is the Portland cement although small amounts of alkali may be contributed from other sources such as aggregates, admixtures and the environment (e.g. seawater or groundwater). In the absence of reactive aggregate the quantity of alkalis in the pore solution will remain reasonably steady with time and some slight increase in concentration may even be observed as part of the pore water is consumed by hydration. In small elements with a high surface to volume ratio some alkalis may be leached from the concrete if the surface is frequently exposed to a supply of moisture. If reactive aggregate is present the alkali in the pore solution and the reactive silica combine to form an alkali-silica gel (this only happens if there is a sufficient concentration of hydroxyl ions to weaken the silica structure).

If the alkali was permanently bound by the reaction product, the alkali (and hydroxyl) ion concentration would eventually decline to a level where the silica was no longer vulnerable to attack. This is clearly not the case as damage due to the reaction continues for many decades in large elements even though the alkalinity of the pore solution appears to have reached equilibrium at a relatively low concentration. The continued reaction can be explained on the basis of the "alkali cycle" shown in Figure 9. This cycle is made possible by the abundance of calcium in the system, which exchanges with some of the alkali (sodium or potassium) in the gel producing a calcium-alkali-silica gel. This releases a portion of the alkalis back to the pore solution where it becomes available to fuel further ASR.

In large concrete elements, such as hydraulic dams, it is conceivable that the alkali-silica reaction will continue until either all the reactive silica or all the available calcium is consumed; calcium hydroxide is likely to be the main source of available calcium. In smaller elements or laboratory specimens the alkali cycle may be interrupted by the continued leaching of alkalis from the specimen.

The use of pozzolans to control ASR is well established and although the primary function of these materials in this role is to sequester alkalis otherwise bound for the pore solution, their ability to combine with lime via the pozzolanic reaction is clearly an additional benefit in terms of minimizing the potential for damaging reaction.

REFERENCES

Bleszynski, R.F. and Thomas, M.D.A. 1998. *Advanced Cement Based Materials*, Vol. 7, pp. 66-78.

Diamond, S. 1983. *Proceedings of the 6th International Conference on Alkalis in Concrete*, (Ed. G.M. Idorn & Steen Rostam), Danish Concrete Association, Copenhagen, pp. 155-166.

Diamond, S. 1989. *Proceedings of the 8th International Conference on Alkali-Aggregate Reaction*, (Ed. K.Okada et al), Kyoto, pp. 83-94.

Diamond, S., Barneyback, R.S. and Struble, L.J. 1981. *Proceedings of the 5th International Conference on Alkali-Aggregate Reaction*, Cape Town, S252/22.

Chatterji, S. 1979. *Cement and Concrete Research*, Vol. 9 (2), 1979, pp. 185-188.

Chatterji, S. and Clausson-Kass, N. F. 1984. *Cement and Concrete Research*, Vol. 14 (6), pp. 816-818.

Hansen, W.C. 1944. *Journal of the American Concrete Institute*, Vol. 15 (3), pp. 213-27.

Kollek, J.J., Varma, S.P. and Zaris, C. 1986. *Proceedings of the 8th International Congress on the Chemistry of Cement*, Vol. 3, Rio de Janeiro, pp. 183-189.

Thaulow, N. and Geiker, M. R. 1992. *Proceedings of the 9th International Conference on Alkali-Aggregate Reaction in Concrete*, (held in London, 1992), Vol. 2 Published by The Concrete Society, Slough, pp. 1050-1058.

Thomas, M.D.A., Nixon, P.J. and Pettifer, K. *Proceedings of the 2nd CANMET/ACI International Conference on Durability of Concrete*, (Ed. V.M. Malhotra), Vol. 2, American Concrete Institute, Detroit, 1991, pp. 919-940.

Thomas, M.D.A. 1996. *Magazine of Concrete Research*, Vol. 48, No. 177, pp. 265-279.

Thomas, M.D.A., Blackwell, B.Q. and Nixon, P.J. 1996. *Magazine of Concrete Research*, Vol 48 (177), 1996, pp 251-264

Wang, H. and Gillott, J.E. 1991. *Cement and Concrete Research*, Vol. 21 (4), pp. 647-654.

FREE LIME CONTENT AND UNSOUNDNESS OF CEMENT

Ivan Odler
Technical University Clausthal
Causthal-Zellerfeld, Germany

ABSTRACT

Free calcium oxide in Portland cement, if present in larger amounts than acceptable, causes an expansion of the hardened cement paste due to its topochemical conversion to calcium hydroxide. The kinetics of this process depends on the burning conditions, in particularly on the maximum burning temperature. The expansive force generated by the $CaO \rightarrow Ca(OH)_2$ conversion may reach a magnitude of ≈150 MPa.

INTRODUCTION

Calcium hydroxide is a regular constituent of Portland clinker. If it is present in small amounts, it does not affect adversely the quality of the cement, however, in larger amounts it may cause expansion and unsoundness. The tendency of free calcium hydroxide to expand is exploited in some expansive cements, but the use of this type of binder is rather limited.

The expansiveness of free lime is the consequence of its hydration to calcium hydroxide

$$CaO + H_2O = Ca(OH)_2,$$

a reaction that is associated with chemical shrinkage of 2.33 percent. Thus, special conditions must exist to cause an external expansion of the cement paste in a process which is not associated with a volume increase of its constituents .

In the presented work we studied the phenomenon of unsoundness caused by the presence of free lime in cement. Particularly, we investigated how the conditions of clinker burning i.e. burning temperature and burning time, affect the kinetics and final extent of expansion caused by the presence of this cement constituent. We will also present our data on the magnitude of the expansive force generated in the hydration of calcium oxide. Finally, we will present our view on why the formation of calcium hydroxide from calcium oxide may or may not be an expansive process, whereas the formation of the same phase during the hydration of dicalcium and tricalcium silicates takes place without any expansion.

EFFECT OF CINKER BURNING CONDITIONS ON EXPANSION

In this series of experiments we used an industrial raw meal of the following oxidic composition: CaO: 69.9%, SiO_2: 19.8%; Al_2O_3: 5.40%; Fe_2O_3: 1.81%; SO_3: 1.18%; MgO: 1.03%; K_2O: 0.91%; Na_2O: 0.40% and TiO_2: 0.24%. To produce clinkers with different free CaO contents, burnt at different burning temperatures, the raw meal, placed in a Pt-dish, was heated at a heating rate of 10°C/min to the selected maximum temperature and then kept at this temperature for different periods of time, before cooling it in air. To obtain free lime contents in the desired range even at 1280°C, 2.0% of SO_3 in form of $CaSO_4$ was added to the raw meal to act as a mineralizer. From all clinkers cements were produced by grinding them with 3.0% SO_3 (in the form of gypsum) to a specific surface area of 300±10 m^2/kg. Also included in the experiment was an industrial Portland cement originated from the same plant. This one was combined with 10 per cent of calcium oxide prepared from $CaCO_3$ (fraction 0.1-0.2 mm) by burning it for one hour either at 900°C or 1100°C.

Pastes of standard consistency produced from these cements were tested for volume stability by the LeChatelier method. The testing apparatus consists of a flexible cylinder mold 30 mm in diameter and 30 mm high, split along its end at one point. The cement paste is placed into the mold and is enclosed at both sides with glass plates. Any expansion that takes place in the paste is magnified by the relative movement of two 150mm long arms soldered to the mold on either side of the split. Molds filled with the paste were placed into water at 20±2°C for up to 28 days. Parallel to it, the same paste was heated up to the boiling temperature and kept at this temperature for 3 hours. At a preset time the distance between the arms was determined as a quantitative measure of the extent of expansion.

Table 1 summarizes the expansion values found after 24 hours and 28 days of curing at 20°C, as well as those found in samples exposed to boiling water temperature right after removal from the water bath. The latter values indicate the ultimate expansion of the paste, as under these experimental conditions the present free calcium oxide undergoes complete hydration. The cement is generally considered to be non-expansive if the distance between the two arms of the LeChatelier apparatus does not exceed 10 mm. Expectedly, among cements burnt at the same maximum temperature the expansion increased with increasing free lime content. It is remarkable, however, that at roughly equal free lime contents the 24 h value depended greatly on the maximum burning temperature and declined distinctly as this temperature increased. In fact, cements burnt at the highest temperature studied, i.e. at 1450°C, exhibited virtually no expansion after 24 hours, even though they expanded distinctly after 28 days. These findings indicate that the reactivity of the free lime present in the clinker depends greatly on the conditions of burning, especially on the maximum burning temperature employed and this determines how fast the cement will expand after mixing with water.

From the Table 1 it may be also seen that additions of calcium oxide produced from $CaCO_3$ by burning at low temperatures did not cause appreciable expansion of the cement paste, even if added in amounts of 10 per cent. This may be explained by the high

Table 1 - Kinetics of expansion caused by free CaO

	CaO_{fr} (%)	Le Chat. 24 h 20°C (mm)	Le Chat. 28 d 20°C (mm)	Le Chat. 3 h 100°C (mm)
1450°C **0'**	**4.4**	**3**	**45 c**	**66 c**
15'	**3.2**	**4**	**28**	**56 c**
30'	**2.9**	**4**	**15**	**32**
1400°C **0'**	**6.2**	**35 c**	-	-
15'	**5.1**	**20**	**44 c**	**59 c**
30'	**4.2**	**8**	**40 c**	**55 c**
1280°C **0'**	**8.4**	**99 c**	-	-
(+2% SO_3) **15'**	**5.8**	**41 c**	-	-
22'	**4.5**	**25**	**40 c**	**45 c**
30'	**3.7**	**15**	**32**	**35**
PC **+ 10% CaO** (900°C)	-	**2**	**2**	**3**
PC **+10% CaO**(1100°C)	-	**3**	**4**	**6**

C = crack formation

reactivity of the CaO that was produced this way, which resulted in complete hydration of this phase prior to setting of the cement paste.

EXPANSION FORCES GENERATED BY LIME EXPANSION

To be able to measure the expansion force associated with the conversion of free CaO to $Ca(OH)_2$ a special apparatus/method was developed:

A mold, shown schematically in Figure 1, was made from a special type of steel with high strength and an exceptionally low deformability under mechanical loading. The mold had space for a test specimen 15x15x60 mm^3 and a steel block 15x25x25 mm^3 (measuring block) made from another type of steel which, unlike that used for the mold, is particularly soft and possesses a relatively high deformability. An intimate contact between the specimen and the measuring block was established by a screw at the opposite end of the mold (not shown in the figure). The expansive force generated in the test specimen produced very small, but well measurable deformations of the measuring block that were measured by a strain gauge attached to its surface. The measuring signal was amplified in an amplifier and recorded in a recorder. The measuring unit had been calibrated by exposing the measuring block/strain gauge assembly to different pressures in a hydraulic press and by measuring the resultant signals.

Figure 2 shows the development of the expansive force found in a paste consisting of 90% C_3S and 10% CaO (fraction 0.1-0.2 mm) mixed at a w/s = 0.40. The CaO had been produced by burning $CaCO_3$ for 6 hours at 1350°C. It may be seen that the expansive force reached a maximum value of about 120 MPa within about 10 days and stayed constant afterwards.

MECHANISM OF EXPANSION

To explain why the formation of calcium hydroxide (portlandite) in the hydration of dicalcium or tricalcium silicate in Portland cement pastes does not cause expansion, whereas the hydration of free lime may, we are offering the following :

In the hydration of dicalcium- and tricalcium silicates the interaction between the mixing water and the compact anhydrous calcium silicate takes place on the surface of the latter. In this reaction, besides of C-S-H, free Ca^{2+} and OH^- ions are formed and enter the liquid phase. Only after reaching a sufficient degree of super -saturation, the barrier to nucleation of portlandite is overcome and this phase precipitates randomly from the liquid phase, without exerting pressure on the surrounding solids (Figure3). Thus, this "through solution" process is not associated with an increase of the external volume of the material.

The texture of free calcium oxide present in Portland clinker differs fundamentally from that of the calcium silicates. This phase forms in the thermal dissociation of calcium carbonate, in which 44 per cent of the original solid converts to gaseous CO_2. Thus, unlike the calcium silicates, calcium oxide possesses a very large internal porosity (about 72 vol. %) and an internal surface that exceeds significantly its external surface. Upon mixing the cement with water, water migrates into the pore space within the free lime particle and here this phase becomes supersaturated with respect to $Ca(OH)_2$ well before significant amounts of calcium oxide dissolve in the bulk liquid. Consequently, crystalline portlandite nucleates and is formed preferentially within the pores of the CaO particle. This, in course, generates an internal pressure and ultimately an increase of the external volume of the particle, as it gradually converts completely

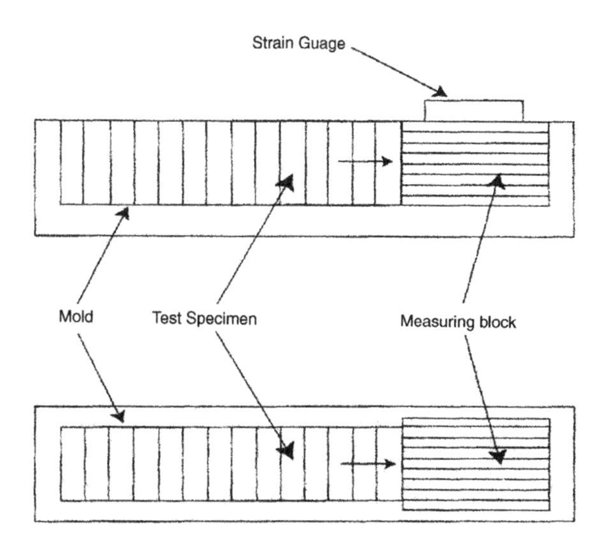

Fig. 1. Assembly for measurement of the expansion force in cement pastes

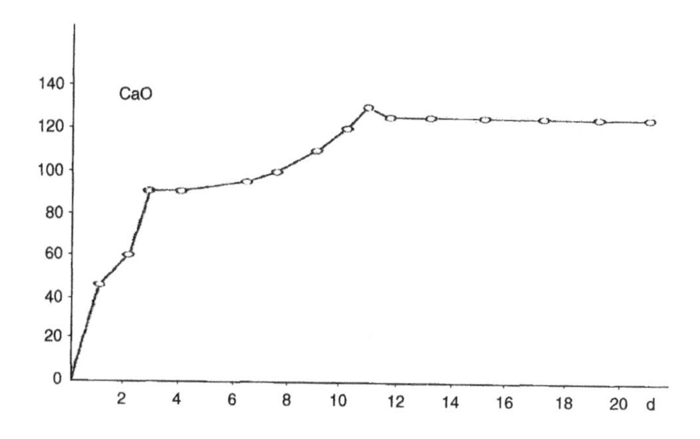

Fig. 2. Development of the expansion force in a paste consisting of 90% of C_3S and 10% of CaO

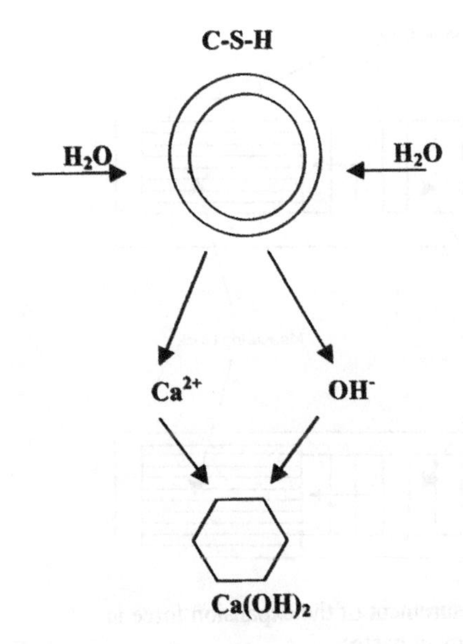

Fig. 3. Schematic presentation of portlandite formation in the hydration of C_3S

from CaO to $Ca(OH)_2$ (see Figure 4). As long as such local volume increase takes place prior to setting and the paste is still of plastic consistency, it may adjust to the existing volume changes by plastic deformation and no harm occurs. On the other hand, if this "topochemical" CaO to $Ca(OH)_2$ conversion takes place in the cement paste after setting, the process generates stresses within the material, which may cause an expansion of the external volume and even crack formation. In industrial Portland cements the free calcium oxide had been formed at high burning temperatures and possesses a very low reactivity. Thus, during hydration, virtually all of it hydrates only after setting, causing the above problems if present in amounts exceeding an acceptable limit.

CONCLUSIONS

It is hypothesized that the expansion of hardened Portland cement pastes caused by the presence of free lime in the binder is the consequence of a delayed topochemical hydration of CaO taking place within the pores of the free lime particles.

The kinetics of expansion associated with the CaO → $Ca(OH)_2$ conversion depends on the burning temperature used in the production of the clinker and declines as

Fig. 4. Schematic presentation of portlandite formation in free lime hydration

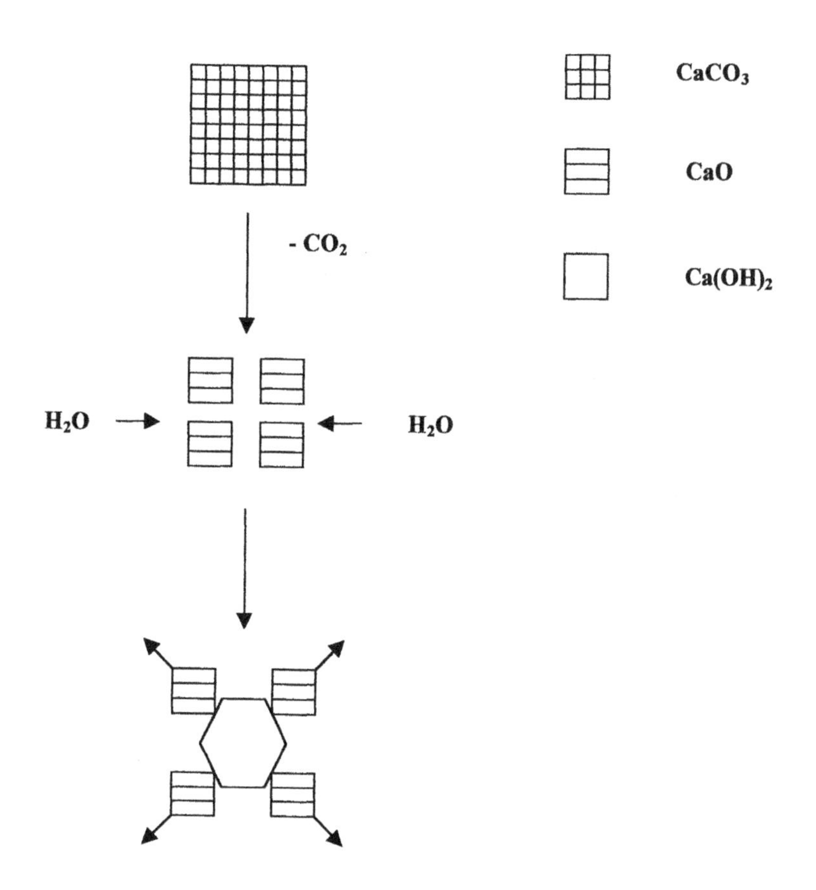

this temperature increases. At equal burning temperatures the final expansion increases with increasing CaO content in cement. It declines, however, with decreasing burning temperature.

The expansive force generated by the conversion increases initially, until it levels off at a constant level. The maximum value may reach a magnitude of 150 Mpa.

DINNER SPEECH

LOOKING FOR SOUL IN CONCRETE

Guest Speaker: **Dr. Robert H. Davis**, Okemos, Michigan

When Jan Skalny first told me he and his colleagues were planning an international workshop for chemists and physicists in Holmes Beach, the plan sounded a little bizarre to me. Holmes Beach is not exactly a scientific Mecca. Then when he told me you were going to talk about calcium hydroxide and concrete, that did it. I doubt that those five words *calcium, hydroxide, concrete, and Holmes Beach* have ever appeared together since the invention of the printing press.

Now you all may have considered Jan's plan perfectly sane. Like almost everyone else in this world, you had probably never heard of Holmes Beach. So, for all you knew, there might be a great university or research center hidden away on Anna Maria Island. But for me there is no such excuse. I not only know Holmes Beach, I am a psychologist and presumably I'm especially qualified to recognize delusions and hallucinations when I see them.

I didn't accept Jan's invitation right off the bat. I told him the story about Groucho Marx, who was once invited to become a member of the exclusive Beverly Hills Country Club. After thinking about it for only a few minutes, Groucho wrote a rejection to the Board of the Club. The note said: *"I wouldn't want to belong to any club that would have me as a member."* Well, I wasn't sure I should speak to any group of hard headed scientists on Anna Maria Island who wanted me to talk to them.

Jan assured me that he wasn't crazy. It was his wife, Magda, who convinced him that you all needed a break. She had heard me give a talk about soul -- that's S-O-U-L -- and she thought it would be a nice "time-out" for all of you to hear something different. Well, I've read the topics in your program and I can tell you up front that this talk will be different.

I begin with a very ancient idea that goes back in the West to the Renaissance and the Neoplatonist philosophers, who were actually psychologists, except there was no such thing in those days. These philosophers said the entire world had *soul*, including all animate and inanimate objects, like buildings and concrete. The term they used for an ensouled world was: *Anima Mundi.* Loosely translated, anima mundi means *world soul.* There's something else you need to know about *anima mundi*, because it's a big part of what I have to say. Anima is also a feminine principle. The male word is *animus.*

So, these renaissance philosophers spoke of a world ensouled, and that's what I'm going to talk about. And now, the secret is out. Both of us, Jan and I, are a little crazy. In fairness I should tell you that Jan may have had some second thoughts about my speech because he called me shortly after inviting me to speak and asked if I would try to make my talk a little bit funny. I told him I didn't have to try. A psychologist talking about soul to a group of concrete specialists is funny enough already.

To really understand what I mean by *soul*, you're going to have forget everything you think you already know about it. First, when I use the word "soul", I don't mean something that leaves the body when you die. I have no idea what happens after you die. I feel like Woody Allen felt about death. *"I don't mind dying," he said. I just don't want to be there when it happens."* And I don't want even to talk about it.

And when I use the word soul, I am not talking about religion. Only politicians seem to want to talk about religion these days.

The key to understanding what I mean by soul is to think in terms of people, objects, and events that stimulate the imagination. Soul is about a particular perspective, a way of looking at the world that elicits fantasy and imagination and ultimately creativity. Two weeks ago, for example, this conference was just a coming event marked on your calendar. But between the day you noted this event on your calendar and this night, something has happened to fill that open space on a calendar with imagination. This conference was not just about chemicals and concrete. It's about the people, the beach, the village, and how they have stimulated your imagination and your creativity. The event has been turned into an experience, and that is what soul is about.

As chemists and physicists you may know that Sir Isaac Newton practiced alchemy his entire life. Insofar as I know, he didn't discover the philosopher's stone, but he did "project" into those mysterious chemical reactions his own imagination and imagination is the key to turning around our thinking and opening us up to new possibilities -- to our creativity.

I'll bet you all had chemistry sets when you were kids. You were probably interested in learning about the periodic table or maybe the loss of alkalinity in a cement paste matrix. Me, I just wanted to see what happened when you mixed different chemicals together. I never succeeded in creating a really giant explosion, which was my primary objective. (There was no Internet with recipes for creating bombs in those days.) But I did get some wonderful and beautiful effects. I was never going to win a Nobel prize in chemistry, but Like Sir Isaac, my imagination was greatly stimulated.

I'd like next to tell you, first, what I mean by soul and why I call it a feminine principle. And second explain how we give soul to "dead" objects and events by personifying and pathologizing them. Just in case you didn't get those two words: ***Personalizing*** and ***Pathologizing***. I'll explain what they mean shortly.

When I use the word soul, I'm using it in the same way it was first used in this country, as far as I know, by the African Americans or blacks. They talk a lot about soul food, and soul music. Nowadays we talk about soul mates. Some folks say that Bush may not be able to pronounce "boutique", but he has soul. Soul is not about head -- that's Gore; its about heart, a word Bush uses a lot and seems to mean. Soul has to do with fantasy and the imagination, which fill events.

Soul means something like deep feelings that cannot be put into words. Soul music you understand with you heart, not your head. It's like love. Think about it this way. You send your wife a Valentine's day card. She opens it expectantly. It says: I LOVE YOU WITH ALL MY HEAD. You what? No, we say HEART, because we know there are things in life beyond the head, beyond words, something more important.

Soul is about the depths, not about heights. It's of the earth -- about people, and children, and life as it is lived. Here are a few slides of some buildings that Jan ordered for me from the Portland Cement Association in Skokie, Illinois.

(Sears Tower, **Chicago, Illinois)**

(United Nations Towers, **Sao Paolo, Brazil, 1997-1998)**

These buildings are not about soul. They're about spirit, soaring heights. Notice the straight lines. Everything is done with a T-square, a ruler, and a compass. These buildings are all masculine, even phallic. Psychologists joke about architects and university presidents having an "Edifice Complex." For years, cities have been caught up in a competition to see which has the biggest building ... Chicago, New York, Hong Kong. They're always at it. Higher and higher. Mine is bigger than yours! Competition. Setting records.

But. soul is not up there. It's down here where the earth is damp. It's about life, and love, and fantasy, and the imagination. And it is often the missing ingredient in aborted efforts to innovate or create something entirely new. It helps to be in touch with your feminine side, if you wish to create. The Greeks knew that the muses were crucial to creativity. More than anything else, soul is about women, who are of the earth.

As scientists, you would, no doubt, like me to give you an operational definition of soul. I've given you all the definition I can. I'm like the judge in my hometown up North who ruled against a man operating a pornography shop and made him close it down. The man's lawyer was incensed. He said to the judge, *"You can't even define pornography."* The judge agreed and then added. *"But I damn well know it when I see it."* People know soul when they see it.

I've said soul is easiest to recognize when something is *personified.* What does *personalizing* mean? Well to *personalize* is to give something a personality -- to give it human qualities, to make it a *person* -- even something inanimate like a car or a building.

You may be a little shocked to hear soul mentioned in connection with buildings and cities. Anthropologists think this is all very primitive -- a form of animistic thinking. Such nonsense, they say to treat an object as if it were living. But the truth is we do it all the time.

Let me see if I can put a little flesh and blood on this idea of ensouling inanimate things. That's a pun. Take this wine. How would an expert describe it? I can almost guarantee that beyond mentioning its acidity, he is not likely to do a chemical analysis. Psychologists have been trying for decades to develop a taxonomy for taste. You know: salty, sweet, bitter, tangy, etc. But do any of those words really describe the delicious taste of this wine? I don't think so.

To adequately describe this wine, you must *personalize* it. That is what experts do. They give it a personality or soul, because there is something about wine that cannot be captured any other way.

First of all, think about the way wine must be served. We have ritualized the service, which is a way of showing our respect for the wine. The bottle must be opened properly. The host must be given an opportunity to taste it -- to make sure it is the right wine, properly aged. It must be served in the correct glass. For red wines a wide rim so that one can enjoy the bouquet, and recall perhaps a trip to Burgundy or the Napa Valley. For champagne a tall glass so that we can enjoy the bubbles and remember anniversaries and weddings.

Now how does an expert describe *the taste* of the wine when he really wants to express his feelings about it. He uses words like these:

AGGRESSIVE/BELLIGERENT
ROBUST
LIVELY
ASSERTIVE
SOFT
SHY/TIMID
PRETENTIOUS

You may smile, but words like these stimulate the imagination. What the expert has done is turn the wine into a person -- he has personalized it. Given it soul.

We do things like this all the time, especially with cars. It's classic in films when some great comedian is shown swearing at his stalled car and calling it names or even kicking it as if it were a person. Only someone like Chaplin would bring flowers to a machine or caress it. And why is it funny? Because we know that it's so ridiculous and yet has a germ of truth. Sometimes we do feel as if our old cars or favorite tools or our homes are almost human. We personalize them.

Children do this quite naturally. The distinction between animate and inanimate, what is ensouled and what is not, is blurred for them.

Many years ago, I asked my four year old grandson what he wanted to be when he grew up. He said he wanted to be a fire truck. Fire truck?

"You mean fireman, don't you?"

"No, I want to be a fire truck."

I've said that soul is not only about personalizing, but it's also about *pathologizing* or what might be called, "Falling Apart". What does that mean?

No one knows why, but we humans go around all the time spontaneously creating illness, morbidity, abnormality, suffering, and disorder in this world. It's as if we have a God-given right to expect the worst. Medical students, for example, are constantly reading about diseases and it is not uncommon for them to imagine they have the diseases they read about.

I have an overweight friend who occasionally goes on a diet. But he always gives up. Finally, I asked him why he couldn't stick with a diet. He told me, that whenever he lost fifteen or twenty pounds, he'd look in the mirror and wonder if he had cancer. To prove to himself that he didn't have cancer, he had to see if he could gain the weight back.

Now *pathologizing* occurs not only with living things, but with inanimate objects as well. When my grandson was about four, he loved the idea of "bad shape". Every once in a while, when we were driving along, he would see something in a state of collapse, like a building or an old car, and he would say, "Wow, that car is in bad shape." Finally, one day I asked him what he meant by "bad shape". He thought about it a long time and finally said. *"They're sick."*

The car is sick. Now think about that. You say what nonsense. Car's don't get sick. But adults *pathologize* inanimate objects all the time. We speak of sick buildings for example. Somehow saying there's a virus lurking in the air conditioning or loose asbestos floating around doesn't describe the seriousness of it. To really capture the essence of what's wrong with that building we project pathology into it. The building is sick. When you really want to describe some thin, anemic looking structure, you might say it's anorexic, and that puts it all in a single word, anorexic. Remember how I described those tall buildings as phallic. I was personalizing.

Now I've said several times that soul has always been about woman. Anima is woman. So is yin in the Chinese symbol of Yin/Yang. From the beginning of time, in all cultures, women are of the earth, of people, of love. Man has his head in the clouds; he soars like the Empire State Building, sometimes right off into outer space.

You remember the wonderful true story of Heloise and Abelard, which tells it all. They had a secret love affair which was discovered by her uncle. She was in love with Abelard, but he was in love with the Church. She was sent in misery to a convent and wrote to him constantly of her love; and he went happily to a monastery and wrote abstract tomes about nominalism. He must have spent some very cold nights there, sleeping on that hard stone bed with a manuscript for a pillow. But that's the way it is with men sometimes. They need someone to remind them to stop and smell the roses, but they are not inclined to listen.

Let's look at a couple of different examples of the difference between men and women and how it relates to the imagination, fantasy, and creativity.

Deborah Tanner, a psychologist, has done a lot of research on the differences between men and women and the way they communicate. From my perspective, her work casts a lot of light on the imagination and the feminine nature of soul.

She's found, for example, that when men talk to one another, they do it to share information. Like at this conference. Or even a telephone call. Here's how it goes.

"Hello."

"Rolf?"

"Yeah."

"Joe."

"Hi."

"Hi."

"The meeting's at nine."

"Good. I'll be there.... Bye"

"Bye."

Maximize the information transfer. Minimize the time wasted talking to another human being. Move on before he starts talking about his personal problems, new cars, painting the house, whatever.

Tanner says women communicate not to pass information but to be with one another. They are talking to have their imagination stimulated. They are looking for soul. Not information. They can happily talk for hours about nothing in particular.

Here's another of Tanner's examples. Has something like this ever happened to you, ladies?

You and your husband are driving along in a strange city and you both realize you are lost. After passing the same gas station for the umpteenth time, you say, *"Let's stop and ask somebody."*

He says, *"Noooo. I can find it."*

Now, this is not just about male ego. It's about not wanting to get trapped in a conversation with another human being. So around you go for a few more times and he finally relents. But, he sends you into the small grocery store to ask directions. You get into a conversation with the lady at the cash register about how bad the tomatoes are this year. You're enjoying soul. You're not thinking about roads that go straight ahead, or turn left then right, etc. You get back into the car and he says, *"Well, what did she say?"* You reply, *"The tomatoes are bad this year because they're coming in from Mexico."*

When men are asked by a stranger for directions, they say something like this. *"You go straight ahead for two miles, or thereabouts, turn right. Go another mile. Right again. You'll see a sign for Highway 365. That's it."*

When the same stranger asks a woman for directions, she says something like this. *"You go up there to where the old widow Jones lives in a yellow house with a dog named, Eddie. There are red geraniums in a window box. Turn there until you come to a corner with an old broken down abandoned wagon. That's where you turn until you see a collapsing barn that the owners don't have the money to fix up -- usually there are kids playing in the yard."* Lots of imagination and soul. How did that wagon get there? Were people in it when it collapsed. Are all those kids from one family? And so on.

In closing, I'd like to say a final word about buildings. Buildings can be feminine as well as masculine. They need not necessarily be hard edged and soar into the sky. They can also be warm and comforting and sheltering. They stimulate the imagination. Buildings with soul are closer to the ground; they're sheltering, with warm colors and curved lines, like a woman's body. When you were a kid did you ever throw a blanket over two chairs and make a tent, a place of the imagination, where you hid away and pretended this was your private place where you could be an Arab or an Indian or anything or anyone you wanted to be. You see how easy children ensoul things -- like a blanket and a couple of chairs.

It's good to personify and pathologize inanimate things, to treat them as if they had souls. There are at least two reasons for this.

First, we tend to respect things with souls more than things that are nothing more than objects. When a forest or a river or a building is just another object or thing, what need is there to respect it? We are not much concerned about abusing and mistreating it. We bulldoze over old buildings without much thought in this country, even when they are filled with imagination and soul. We're told it's cheaper to tear 'em down and replace them than to repair them. Our developers can be as impersonal and destructive as earthquakes in the Middle East. They need a little more soul.

And the second reason for ensouling the world concerns creativity. Fantasy and the imagination, which are so to speak the soul of soul, are essential ingredients for creativity. We do our most creative thinking when we get out of the boxes in which we are trapped, the ordinary way of thinking, and move to something entirely new, unexpected and different--even whimsical. Sound's a bit like coming to Holmes Beach for an international conference about calcium hydroxide and concrete, doesn't it?

I'm going to close by showing you a few pictures that may convince you that even buildings can have soul.

(A. Gaudi, *Casa Batllo,* Barcelona, 1904-1906)

(A. Gaudi, *Casa Batloo – Vestibule and Stairway,* Barcelona, 1904- 1906)

(A. Gaudi, *Crypt of the Guell Estate Church,* Barcelona, 1908-1917)

(Gaudi, *Casa Mila*, Barcelona, 1906-1911)

(Gaudi, *Casa Mila*, Barcelona, 1906-1911)

Jazz at the reception: The Kercher Quintet

Enjoying the reception: Karin Gebauer and Sidney Diamond

Lily Farkas and Vagn Johansen

Ana Hidalgo and Emery Farkas

Aase and Niels Thaulow

Marika and Ivan Odler

... and the WINNERS are:

Hal Taylor and Karen Scrivener

KEYWORD AND AUTHOR INDEX

Printed and bound by CPI Group (UK) Ltd, Croydon, CR0 4YY

16/04/2025

14658450-0001